T0290888

Geotechnics of Venice and Its Lagoon

Without any protective intervention, the historic city of Venice and its surrounding islands would suffer rapid deterioration due to the increased frequency of tidal flooding, as the gap between land surface and sea levels has reduced due to a coupled effect of climate change-induced sea-level rise and natural and anthropic subsidence.

Geotechnics of Venice and Its Lagoon provides a clear and comprehensive illustration of the extensive geotechnical aspects of not only the various environmental problems such as land subsidence and wetland surface reduction but also solutions such as the design of the tilting gate foundations against high tides and the restoration and improvement of the drainage system of the renowned Piazza San Marco, which have been necessary for the preservation of the extraordinary cultural heritage of Venice. Readers will gain a better understanding of the complex phenomena occurring in the sensitive Venice silts, whose hydro-mechanical behavior has required comprehensive laboratory and site investigations and modeling. The book provides:

- An authoritative analysis of one of the largest and most important geotechnical issues in the world;
- A description of a detailed case study of an ongoing engineering solution.

The book will be useful for engineers worldwide and is also an excellent reference for students.

Paolo Simonini is Professor of Geotechnical Engineering and former Dean of the School of Engineering at the University of Padova, Padua, Italy, and a full member of the Italian government's Council for Public Works.

Geotechnics of Venice and Its Lagoon

Paolo Simonini

CRC Press

Taylor & Francis Group

Boca Raton London New York

CRC Press is an imprint of the
Taylor & Francis Group, an **informa** business

Cover image: Paolo Simonini

First edition published 2024
by CRC Press
2385 NW Executive Center Drive, Suite 320, Boca Raton FL 33431

and by CRC Press
4 Park Square, Milton Park, Abingdon, Oxon, OX14 4RN

CRC Press is an imprint of Taylor & Francis Group, LLC

Library of Congress Cataloging-in-Publication Data

Names: Simonini, Paolo, author.
Title: Geotechnics of Venice and its lagoon / Paolo Simonini.
Description: Boca Raton : CRC Press, 2023. | Includes bibliographical references and index.
Identifiers: LCCN 2022059026 | ISBN 9781032049564 (hardback) | ISBN 9781032049588 (paperback) | ISBN 9781003195313 (ebook)
Subjects: LCSH: Geotechnical engineering--Italy--Venice. | Hydraulic engineering--Italy --Venice. | Subsidences (Earth movements)--Italy--Venice. | Venice (Italy)--Threat of destruction. | Venice, Lagoon of (Italy)
Classification: LCC TA705.4.I8 S56 2023 | DDC 624.1/510945311--dc23/eng/20230315
LC record available at https://lccn.loc.gov/2022059026

ISBN: 978-1-032-04956-4 (hbk)
ISBN: 978-1-032-04958-8 (pbk)
ISBN: 978-1-003-19531-3 (ebk)

DOI: 10.1201/9781003195313

Typeset in Sabon
by Deanta Global Publishing Services, Chennai, India

Contents

Preface

Hundreds of books have been written about the extraordinary city of Venice and its lagoon, each of them addressing a different aspect of this complex and fascinating city, but no one concering with geotechnical issues. The city of Venice and the surrounding lagoon have always attracted my interest; first of all from the historical and artistic point of view, and subsequently, year after year, my interest in the relevance of geotechnical engineering for its preservation has grown. During my pluriannual research experience in geotechnical engineering, I have often been consulted for advice regarding the geotechnical aspects of this environment. For example, I have been involved in the mechanical characterization of the Venetian lagoon's subsoil, or in studies and analyses related to interventions protecting against exceptional tides, or in others connected to the preservation of the historical and cultural heritage and to a better characterization of environmental elements.

In this unique city, which stands in such close contact with the sea, geotechnical issues have assumed a significant role. In fact, its historical buildings and extraordinary characteristics, as well as its future existence, strictly depend on the lagoonal environment and its evolution. First, the natural tidal variation of the sea level within the lagoon affects the city of Venice each day, with the margin of security of this rather precarious equilibrium eroding at an ever-increasing rate. The rate of environmental deterioration is being accelerated by the increasingly frequent flooding of the historical city, caused by rising sea levels due to climate change, natural subsidence, and a local man-induced subsidence, brought about particularly between 1946 and 1970. The ever-larger gap between the level of the sea and the surface of the islands has caused a loss in ground elevation of 26 cm from the beginning of the last century. Chapter 1 describes the characteristics of the lagoon environment as well as its tendency to be increasingly flooded.

This phenomenon became particularly clear when, on the 4th of November 1966, an exceptional flood took place, during which the entire city and the surrounding islands were completely submerged by a tide that lasted for many hours.

The impact on the city of the 1966 flood was so momentous that it induced the Government of the Italian Republic to enact a Special Law for

Venice (Law no. 171 of 16 April 1973) aimed at the safeguarding of Venice and its lagoon (*Interventi per la salvaguardia di Venezia/Interventions for the Safeguarding of Venice*). This represents the first organic legislation enacted after the dramatic flood of 1966.

At around the same time, the Italian government, along with the National Research Council, commissioned the Italian Oil Company Agip Mineraria to dig a 950 m deep borehole in the Venice area, which was intended to be a preparatory study on the relevance of the subsidence phenomenon. The sampled sediments, which cover a time interval from over 2 million years ago to today, provided materials for studies that were at the basis of the subsequent, increasingly detailed research on the evolution of the Venetian territory, whose geomorphology and geology are presented in Chapter 2.

The 950 m deep borehole was integrated by two other boreholes, the 120 m deep VE1-bis and then the 400 m deep VE2, to study the aquifer and aquitard system and to measure geotechnical properties useful for an evaluation of the subsidence problem. During the central decades of the last century, the industrial area of Mestre developed on the closest mainland, and consequently, a large-scale water extraction was performed from the 350 m deep aquifers underlying Venice, reducing the piezometric levels in the deep aquifers and generating ground surface settlements measuring up to 150 mm throughout Venice over the same period. Aquifer depletion was halted in 1973 and so stopped the associated ground subsidence as well. By then, however, enormous damage had been done.

Alongside anthropogenic subsidence, slight natural subsidence also takes place. Natural subsidence is caused by a combination of secondary compression, prevalently occurring in the upper soil deposits, as well as by the processes of oxidation of surface peaty soils, which affect especially the marshes and peatlands subjected to tidal cycles. To keep the evolution of subsidence under control, continuous surveying has continued up to nowadays. These processes are both presented in Chapter 3.

Several interventions were planned by the Special Law in Venice, such as restoration of quay walls, raising of the islands' pavements, protection and reconstruction of salt marshes, and, the most important intervention, the enormous project *MoSE* (an acronym for *Modulo Sperimentale Elettromeccanico*, Experimental Electromechanical Module) involving the design and construction of movable gates located at the three lagoon inlets. These gates, controlling the tidal flow, temporarily separate the lagoon from the sea at the occurrence of particularly high tides, thereby protecting the historic city.

New geotechnical investigations were therefore carried out to characterize Venetian soils at the inlets to achieve a suitable design of the movable gate foundations. The Geotechnical Group of the University of Padova, and especially myself, have been involved in planning a small but meaningful part of these investigations, which were based on the recent advancements in geotechnical laboratory and site testing.

From the geological and geotechnical investigations carried out, it has been found that the lagoon soils are characterized by a predominant silt fraction, combined with clay and/or sand. These form a chaotic interbedding of different sediments, whose basic mineralogical characteristics vary slightly, as a result of similar geological origins and common depositional environment.

This latter feature, together with the relevant heterogeneity of soil layering, suggested that research concentrate on some selected test sites, considered representative of typical soil profiles, where relevant in situ and laboratory investigations could be carried out for a careful characterization of the Venetian lagoon soils.

The first test site, namely the *Malamocco test site*, was determined in the early 1990s at the Malamocco inlet. Within a limited area, a series of advanced site investigations that included boreholes with undisturbed sampling and laboratory testing, seismic piezocone, dilatometer, self-boring pressuremeter, and cross-hole tests were performed on contiguous verticals. In addition, a continuous borehole was carried out for very careful soil mineralogical classification.

The comprehensive laboratory test program completed at the *Malamocco test site* emphasized the very heterogeneous, high silty content and low-structured nature of the Venetian soils, emphasizing the difficulty to characterize even the simplest mechanical properties with a certain degree of accuracy only from geotechnical laboratory tests.

In the early 2000s, a second test site was therefore selected, namely the *Treporti test site*, which was located at the inner border of the lagoon, very close to the Lido inlet. The goal of this new site was to measure directly in situ the stress–strain–time properties of the heterogeneous Venetian soils. At the *Treporti test site*, a vertically walled circular embankment was constructed, thus measuring, along with and after the construction, the relevant ground displacements together with the pore pressure evolution.

To this end, the ground beneath the embankment was heavily instrumented using plate extensometers, differential micrometers, GPS, inclinometers, piezometers, and load cells. Boreholes with undisturbed sampling and traditional and special dilatometer and piezocone tests were employed to characterize soil profile and estimate the soil properties for comparison with those directly measured in situ.

The results of this second special investigation were particularly relevant to accurately calibrate site testing techniques, to measure relevant soil properties, and to formulate, calibrate, and propose new constitutive models.

The main outcomes of these two research projects are presented in Chapters 4 and 5.

The MoSE barrier system consists of four rows of fold-away gates constructed at the lagoon inlets. Made of steel, these oscillating buoyancy flap gates are raised from the sea floor to prevent water from entering the Venice lagoon when high tide is forecast. When inactive, the floodgates are folded and embodied in concrete caissons buried at the bottom of the inlet channels, so that they are completely invisible. The geotechnical design and the construction of the barriers posed countless challenges, the most difficult being those related to the prediction of the caisson settlements lying on settlement reducing piles. The prediction and control of total and differential settlements were particularly complex and were regarded as the most crucial aspect in geotechnical design due to the presence of large rubber joints, which needed to provide waterproof contact between two adjacent caissons, thus guaranteeing accessibility of the service tunnels for maintenance of the electromechanical equipment operating the mobile gates. Chapter 6 briefly summarizes these aspects.

Chapter 7 deals with the preservation of the landscape typifying the tidal environment of the Venice lagoon, which is composed of marshes and wetlands largely covered by halophytic vegetation and subjected to tidal fluctuation, with periods of submersion depending on tidal cycle amplitude and local topography. The eustatic rise in sea level, subsidence, and increasing anthropic pressure have caused considerable environmental damage to the lagoon ecosystem, including relevant changes to the sediment balance of the basin and the vanishing of large marsh areas. To study the complex interaction between marshes and wetlands and the surrounding environment, the upper marsh soil has been both site and laboratory investigated and then instrumented to monitor the groundwater pressure as a function of tide excursion. More recently, a special loading system has been designed and used to measure the compressibility of upper marshy soil, with the aim of modeling their long-term settlements.

Chapters 8 and 9 concern the preservation of the cultural heritage of the historic city.

Chapter 8 describes the typical footings of historic Venetian buildings and the possible effects of their degradation. The buildings' foundations rest on wooden planks or short, small-in-diameter wooden piles, embedded at a very small depth. Modest in size and closely spaced wooden piles help to improve the mechanical behavior of the soft clayey silt, which characterizes the shallowest layer of the Venice lagoon, thus reducing the expected settlements. Wooden piles were long believed to last indefinitely, as they are permanently waterlogged. Disproving this assumption, recent evidence shows that anoxic bacteria can seriously deteriorate wood even in anoxic conditions. The effects of wood deterioration on the mechanical behavior of the foundation over time were investigated and modeled by numerical analysis, showing that wood degradation leads to stress transferring from pile to soil causing an increase in settlements.

Chapter 9 describes the condition of Piazza San Marco, the most famous square of the city, which is located on the lowest-elevated island forming the city and therefore is often flooded during very high tide events. This is the most recent geotechnical study carried out in the Venetian area.

In order to design cost-effective and non-intrusive protection interventions for Piazza San Marco, a deep understanding of flooding mechanisms and the relationship between groundwater pressure and tidal oscillations was necessary. Chapter 9 presents and discusses the results of geotechnical surveys and measurements of pore pressure oscillations in the soil of Piazza San Marco. The results provided important information that guided the design of the project and could be of interest to similar coastal areas. Results showed that significant pressure oscillations occur in the subsoil and should not be neglected when stabilizing horizontal architectural structures, such as historical mosaics and paving.

All these diverse experiences emerge to provide an overall comprehension of the geotechnical issues related to this extraordinary city, inserted in its equally extraordinary lagoon environment. The information presented here may not be complete nor exhaustive but can provide a good starting point, which I gladly share with you. Chapter 10 contains some personal considerations I have developed thanks to over 25 years of studying Venetian soils.

Padova, December 2022

(a)

(b)

Figure 0.1 (a) and (b) Dramatic views of the Riva degli Schiavoni and the Doge's Palace at the occurrence of an extreme tide in November 1966. Credit: Cameraphoto Epoche, Archivio Storico Comunale – Celestia.

Acknowledgments

I wish to express my warmest thanks to my family, Simonetta, Elena and Silvia, for lovingly supporting me and my work, and to all the friends and colleagues whose contributions were fundamental to write this book and to the significant advancements in understanding the complex hydro-mechanical behavior of these non-textbook soils:

- First of all, my colleague Simonetta Cola, full professor of Geotechnics at the University of Padova; the early studies on the Venice soils and most of the contents of this book are entirely due to the research carried out together, continuously discussing how to suitably characterize these non-textbook soils
- Giuseppe Ricceri, formerly professor and head of the Geotechnical Group at the University of Padova for providing grants to investigate the Venetian ground and stimulating discussions on soil behavior aimed at designing the tilting gate foundations, in which he was involved
- Francesca Ceccato, senior lecturer at the University of Padova, who prepared her master's thesis on the behavior of wooden piles as improving elements and, ten years later, investigating with me relevant issues concerning the tide-induced evolution of pore pressure in the Piazza San Marco
- Giorgia Dalla Santa, formerly at the Consorzio Venezia Nuova, now research associate at the University of Padova, who helped me to integrate all the contributions in a unique framework producing the final version of the book
- Sandra Donnici and Luigi Tosi of CNR ISMAR Venezia, who helped me in understanding the complex geological history of the lagoon as well as the subsidence evolution along with the centuries;
- Valentina Berengo, who developed the PhD thesis on the time-dependent behavior of Venetian silts
- Veronica Girardi, formerly PhD student at the University of Padova;
- All the technicians of the Geotechnical Laboratory of the University of Padova, Oscar Garbo, Antonio Gobbato, and Mattia Donà

- Guido Gottardi and Laura Tonni, University of Bologna, Italy
- Michele Jamiolkowski, Emeritus Professor at the Technical University of Torino, Italy
- Silvano and Diego Marchetti, Paola Monaco, University of L'Aquila, Italy
- Sara Amoroso, University of Chieti, Italy
- Helmut Schweiger and Franz Tschuchnigg, Technical University of Graz, Austria
- Juan Pestana, Emeritus Professor at UC Berkeley, and Geosyntec Consultants, Boston, USA
- Giovanna Biscontin, National Science Foundation, USA
- Guido Biscontin, formerly at the University of Venice
- Roy Butterfield, Emeritus Professor at the University of Southampton, UK
- Martino Leoni, Peter Vermeer, and Thomas Benz, formerly University of Stuttgart, DE
- Marco Uzielli, University of Florence, Italy
- Lorenzo Sanavia, Pietro Teatini, and Claudia Zoccarato, University of Padova, Italy
- Consorzio Venezia Nuova, Italy
- Magistrato alle Acque Venezia, now Provveditorato Interregionale per le Opere, Italy
- Pubbliche per il Veneto, Trentino Alto Adige e Friuli Venezia Giulia, Italy
- Soprintendenza di Venezia, Italy
- CORILA, Italy

Chapter 1

The recurrent flooding of the lagoon city

1.1 INTRODUCTION

The Lagoon of Venice is a humid area measuring about 550 km², situated in Northeast Italy along the coast of the Adriatic Sea. The world-renowned historic city of Venice, together with surrounding islands such as Burano, Murano, Torcello, Sant'Erasmo, and others, covers only 8% of the area. The lagoon is connected to the Adriatic Sea by three inlets of varying dimensions and depths. The water level has an average depth of about 1 m, which fluctuates according to the tides, but the morphology of the lagoon bottom varies from tidal flats to excavated channels and shallow areas. Figure 1.1 shows a global view of the Venice lagoon.

The historical development, the extraordinary characteristics, and the future existence of the world-famous historic city of Venice, the former capital of the Serenissima Republic of Venice, are closely linked to the mirror of shallow waters in the middle of which the treasured city rises. The lagoon is also an extraordinary natural environment composed of salt marshes, tidal flats, reedbeds and channels, and rare underwater and abovewater ecosystems, populated by numerous species of plants and animals, some of which are protected under European regulations.

Coastal lagoons are transitional environments that develop where land and sea converge and are in continuous and natural transformation due to the dynamic interaction between erosive and sedimentary processes related to tidal currents, wave energy, flows of suspended solids, and the presence of vegetation that stabilizes the soil, as well as changes in relative sea level and subsidence. The lagoons are, on a geological scale, by nature short-lived formations (see Chapter 2 for more details). Naturally, they tend to either silt up, in the case where sedimentation processes of solid materials brought by rivers prevail, or evolve into a branch of the sea in the case where processes of erosion, subsidence, and eustacy predominate (Costanza et al. 1997; Perillo et al. 2018). Thus, although lagoons are ecosystems in continuous transition over time and space, the Venice lagoon has been carefully preserved with great wisdom by means of constant human interventions over the centuries. Venetian citizens have looked after their own

DOI: 10.1201/9781003195313-1

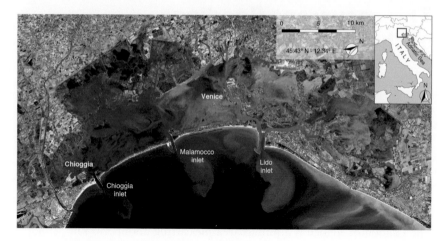

Figure 1.1 **A global view of the Venice lagoon (modified from Copernicus Sentinel Data** 2020, https://scihub.copernicus.eu/).

socio-economic interests by reinforcing the islands, dredging navigation channels, protecting the sea inlets, and even diverting the principal rivers that flow into the lagoon (D'Alpaos 2010).

1.2 THE TIDE IN THE LAGOON

The three inlets that connect the Venice lagoon to the Adriatic Sea allow the continuous inflow and outflow of seawater. Here, the tide is semi-diurnal; that is, two high and two low tides with almost equal amplitude enter and exit the inlets daily, and each time replenish the water in all areas of the lagoon. In general, the two most important driving forces of the tide are astronomical and meteorological.

On a planetary scale, tides are periodic rises and falls of the sea surface caused by the combined gravitational attraction of the Moon and the Sun and the centrifugal force due to the rotation of the Earth–Moon system (astronomical tide). The magnitude of the astronomical tide depends on the moon phase: when the sun and the moon are aligned with each other, in full and new moon, the high and low excursions of the tide are more pronounced (called syzygial tides). When the moon is in the middle of its waning or rising cycle, the tides are less extensive and sometimes present only one daily maximum and minimum, with tidal currents almost non-existent (quadrature tides). Secondarily, the tidal level is influenced by the morphological characteristics of the site, such as the depth of the seabed or the shape and size of the body of water delimited by the coasts, as well as by meteorological disturbances.

In the Northern Adriatic Sea, the main meteorological factors affecting the tide are the local atmospheric pressure, winds (blowing both locally and in the whole Adriatic region), and the quantity of water discharged by rivers into the lagoon.

Low local atmospheric pressure is associated with higher tides, also because it is usually combined with heavy rainfall directly in the lagoon area as well as in the lagoon catchment area, resulting in greater contributions from the inflowing rivers. In addition, differences in atmospheric pressure over the Northern Adriatic (low pressure) and Southern Adriatic (high pressure) can also temporarily create a decimeter-large difference in sea level.

As shown in Figure 1.2, in the Venice lagoon, the local prevailing winds are the Sirocco wind and the Grecale wind. The Sirocco blows from the southeast, following the major axis of the Adriatic Sea, while the Grecale wind, blowing from the northeast and known locally as the Bora wind, follows the major axis of the lagoon itself (Vilibić et al. 2017).

A strong and prolonged Sirocco wind, for example, can push large amounts of water toward the Northern Adriatic Sea, temporarily raising the level by as much as 1 m in extreme cases and hindering the outflow of water from the lagoon toward the sea at low tide, thus lengthening the high tide. The Bora wind, which often blows impetuously, alters the stability of the water level inside the lagoon, facilitating the accumulation of water in the southern part and also hindering outflow into the sea.

According to the entity of the components involved in the phenomenon, the tide can present very different amplitudes. In particular, the concurrence of different factors can raise the water level inside the lagoon in conditions

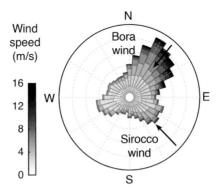

Figure 1.2 Wind statistics for the period 2000–2020 (original elaboration from datasets in https://www.comune.venezia.it/content/dati-dallestazioni-rilevamento), highlighting the two morphologically relevant wind conditions (northeasterly Bora wind and southeasterly Sirocco wind).

of poor outflow, leading to very high tides. Historically, there are known occasions in which Sirocco and Bora winds have acted simultaneously, generating exceptionally high waters.

Through scientific analysis of the astronomical factors, the astronomical tide level may be predicted over the long term. The meteorological tide components, on the other hand, can greatly influence the tidal excursion and can be forecasted with quite high reliability but with little notice (Centro previsione e segnalazione maree – Comune di Venezia, center for tide forecasting and alerting, Municipality of Venice). Based on the analysis of these meteorological phenomena, it is possible to predict the tide level for the current day and the following two days. Figure 1.3 reports an example of a tidal forecast.

In the Venice lagoon, the tide is characterized by a slight excursion, which ranges from 0.4 m to 0.6 m and rarely exceeds 1.6 m during high tides. In addition, as dozens of tidal gauges spread out in the lagoon have registered, the tidal excursion varies to some extent in different locations of the lagoon. Referring to the tide measured at the inlets, the gauges have registered 2.5-hour-long lag times and up to 5% reduction in the peaks (Di Nunno et al. 2021).

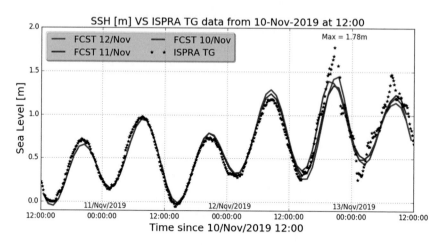

Figure 1.3 Predicted sea levels in Venice during the mid-November 2019 *acqua alta* events. Red line: *MedFS* forecast produced on 10 November for the following three days. Blue line: *MED MFC* forecast recomputed on 11 November. Green line: *MED MFC* forecast recomputed on 12 November. These are compared to observations (black dots) at *ISPRA* Tide Gauge. Time-series from 10 November 2019 at 12:00. Source: *Copernicus Marine MED MFC* (see https://marine.copernicus.eu/news/record-venice-acqua-alta-copernicus-supports-flood-monitoring-and-storm-surge-forecasts).

1.3 TIDAL EFFECTS ON THE LAGOON ENVIRONMENT AND COUNTERMEASURES TAKEN

Clearly, the entire lagoon and the city of Venice are affected by tidal excursions. From the ecosystemic and environmental points of view, extremely high tides and storm surges, by increasing wave resuspension and submersion periods, may represent an important geomorphic driver of salt marsh evolution and essential suppliers of sediment (Tognin et al. 2021). In fact, the salt marshes are in a fragile equilibrium with the tides and their variations because the marshes are intertidal environments that exist in an extremely narrow altimetric interval (see Chapter 7 for more detail); thus, submersion during high tides promotes vertical accretion by sedimentation of the suspended sediments.

On the other hand, the historical city of Venice and its treasures suffer when high tides cause waters to rise, flooding urban areas; the percentage of the urban areas that are flooded depends on the local altimetry and changes in the level of the tide. When tides exceed +1.0 m Piazza San Marco, the area adjacent to the Palazzo del Doge known as 'Piazzetta San Marco' and the quay facing the lagoon, which are all located in one of the lowest parts of the city, are flooded. The tide level is commonly referred to as the mean sea level which, in the case of the Venice lagoon, is measured at the mareograph located in front of the Basilica of Santa Maria della Salute (s.l.P.S. = sea level at 'Punta della Salute') (data from the meteorological station at the 'Punta della Salute' are discussed in the following chapters). The mean sea level measured here is 23.57 cm lower than the actual Italian mean sea level, set in 1942.

When tides exceed +1.2 m, 40% of the city area is flooded, while tides of +1.4 m above s.l.P.S. flood more than 60% of the city, as shown in Figure 1.4.

Increased tidal flooding over time is presenting a genuine threat to the city of Venice; furthermore, the phenomenon is rapidly worsening as a consequence of the coupled effect of subsidence and eustatic rise of sea level due to climatic change, as described in Chapters 2 and 3. Considering only the tide exceeding 1.10 m (commonly referred to as the 110 cm tide), in 100 years the number of exceptional tides per year has increased from 1 to more than 20 (Figure 1.5).

High waters jeopardize life in the city, cause serious inconvenience for the population and businesses, and lead to a slow but irreversible deterioration of physical structures and of the unique artistic and architectural heritage. For this reason, after the aforementioned dramatic events caused by the exceptionally high tide on 4 November 1966, when the water reached a level of 194 cm over the s.l.P.S., the Italian State enacted the *Legge Speciale per Venezia* – Special Law for Venice (Laws 17/1973, 798/1984, 360/1991 and 139/1992). This law stipulates the definition of an integrated plan of

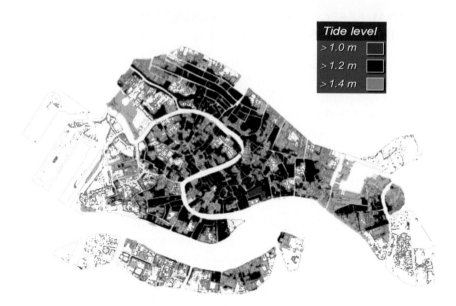

Figure 1.4 Flooded areas in the city of Venice as a function of tide level (from *Consorzio Venezia Nuova – Magistrato alle Acque*).

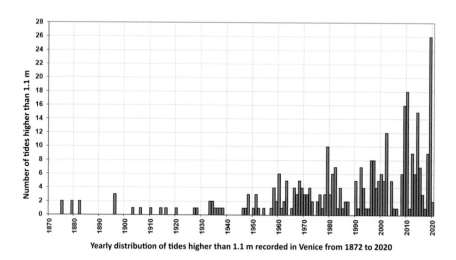

Yearly distribution of tides higher than 1.1 m recorded in Venice from 1872 to 2020

Figure 1.5 Evolution of the number of tides exceeding 1.10 m above m.l.s. over the years. Source: *Centro Previsioni e Segnalazioni Maree – Comune di Venezia, Center for tide forecasting and alerting, Municipality of Venice.*

interventions that combines defense from floods and environmental measures involving the city of Venice and the entire lagoon. This Special Law
was followed by other legislative measures that defined the strategic objec-
tives, the procedures needed in order to achieve them, and the responsibili-
ties of the various institutional subjects involved.

The intervention plan has grown over time and has included the following:

- The reinforcement of the littoral by means of coastal nourishment
  and recalibration of outer breakwaters, to contrast the erosion of the
  littorals in order to defend the area from sea storms;
- The construction of mobile barriers, known as the MoSE (an acronym for *Modulo Sperimentale Elettromeccanico*, Experimental
  Electromechanical Module) system, at the three lagoon inlets, to
  block high tides (higher than 110 cm on the Punta della Salute datum)
  from entering the lagoon (see Chapter 6);
- Interventions leading to the protection and reconstruction of salt
  marshes and wetland habitats (see Chapter 7);
- Local defense of the inhabited areas to 'raise' quaysides and paved
  public spaces in the lowest parts of the urban areas up to a level
  higher than 110 cm, and specific interventions to protect particularly
  important areas of the city such as Piazza San Marco, as reported in
  Chapter 9;
- Finally, interventions aimed at sealing off polluted sites and improving the quality of water and sediments in the whole lagoon, such as
  renaturation of reclaimed lagoon areas and seabed elevation to reduce
  wave motion (see https://www.mosevenezia.eu/).

To correctly manage the high tides and regulate the mobile barriers installed
at the inlets, a tide forecast center was established by the Venice Municipality
in 1980 (ICPSM – *Istituzione Centro Previsione e Segnalazione Maree,
setting up of the center for tide forecasting and alerting, Municipality of
Venice*), which developed a set of numerical models to predict a storm surge
from the astronomical prediction of the tide and measured and predicted
water levels and meteorological data (Mel et al. 2014).

The forecasting of high tides is a crucial point in the decision-making
process that dictates when the mobile barriers at the inlets should be raised.
This is decided according to specific 'operating rules', since activation of the
barriers must be decided roughly three hours before the estimated peak. In
fact, the tides potentially exceeding the threshold are identified five days in
advance and are then monitored by continuously verifying the tide forecast
with real-time data in order to correctly predict the expected water level
and determine whether or not to activate barrier closure.

REFERENCES

Centro Previsioni e Segnalazioni Maree – Comune di Venezia. https://www.comune .venezia.it/it/content/centro-previsioni-e-segnalazioni-maree (accessed March 22, 2022).

Copernicus Marine MED MFC. https://marine.copernicus.eu/news/record-venice -acqua-alta-copernicus-supports-flood-monitoring-and-storm-surge-fore-casts (accessed May 20, 2022).

Costanza, R., D'Arge, R., De Groot, R.,Farber, S., Grasso, M., Hannon, B., Limburg, K., Naeem, S., O'Neill, R.V., Paruelo, J., Raskin, R. G., Sutton, P., and Van den Belt M.1997. The value of the world's ecosystem services and natural capital. *Nature* 387(6630), 253–260.

Di Nunno, F., De Marinis, G., Gargano, R. and Granata, F. 2021. Tide prediction in the Venice Lagoon using Nonlinear Autoregressive Exogenous (NARX) neural network. *Water* 13(9), 1173.

D'Alpaos L. 2010. L'evoluzione morfologica della Laguna di Venezia attraverso la lettura di alcune mappe storiche e delle sue carte idrografiche. Comune di Venezia - Istituzione Centro Previsioni e Segnalazioni Maree, Legge Speciale per Venezia. Ed. Europrint (TV) (*in Italian*).

Mel, R., Viero, D. P., Carniello, L., Defina, A. and D'Alpaos, L. 2014. Simplified methods for real-time prediction of storm surge uncertainty: The city of Venice case study. *Advances in Water Resources* 71, 177–185.

MoSe Venezia – Official site. https://www.mosevenezia.eu/.

Perillo, G., Wolanski, E., Cahoon, D. R. and Hopkinson, C. S. 2018. *Coastal Wetlands: An Integrated Ecosystem Approach*. Amsterdam, Netherlands: Elsevier.

Tognin, D., D'Alpaos, A., Marani, M. and Carniello, L. 2021. Marsh resilience to sea-level rise reduced by storm-surge barriers in the Venice Lagoon. *Nature Geoscience*, 14(12), 906–911.

Vilibić, I., Šepić, J., Pasarić, M. and Orlić, M. 2017. The Adriatic Sea: A long-standing laboratory for sea level studies. *Pure and Applied Geophysics* 174(10), 3765–3811.

Chapter 2

Geology of the lagoon

2.1 INTRODUCTION

The entire geological history of the Venetian territory was deeply investigated through the VE1 sediment core, extracted from the subsoil of Isola Nova del Tronchetto in spring 1971. The sediments sampled are 950 m thick and cover the time interval from over 2 million years ago to today.

The VE 1 borehole was carried out by the Italian oil company *Agip Mineraria* (Figures 2.1 and 2.2) and was intended as a preparatory study of the subsidence phenomenon, which was particularly relevant in the last century. The study of the borehole samples also helped to plan interventions for safeguarding the lagoon, which were then beginning to be addressed and discussed. The results of that operation provided the soil samples for the studies that were the basis of the subsequent, increasingly detailed research on the evolution of the Venetian territory.

While VE1 borehole tells the more distant past of this area, the shallower sediments record its most recent evolution, from thousands of years ago to the most recent centuries.

From the first VE1 borehole, it was determined that lagoon deposits are characterized mainly by the presence of an important silt fraction, combined with clay and/or sand. These form an interbedding of different sediments, whose basic mineralogical characteristics vary relatively slightly as a result of similar geological origins and a common depositional environment.

2.2 STRATIGRAPHIC AND CHRONOLOGICAL FRAMEWORK

The area of the Venice lagoon is located in the foreland basin between the southern Alps to the north and the Apennines to the southwest (Carminati et al. 2003), on the Adriatic lithospheric plate. The curvature of the monocline linked to the subduction of the Adriatic plate below the Apennines gradually decreases from the Po River basin toward the northeast, starting

DOI: 10.1201/9781003195313-2

Figure 2.1 The drilling tower of the VE1 well in May 1971, "Isola Nuova del Tronchetto" (Venice) (photo Serandrei-Barbero).

Figure 2.2 Worksite operations for sediment recovery during the drilling of the VE1 borehole execution (photo Serandrei-Barbero).

from values of about 20° up to values close to 0°. Venice is located on a segment of the monocline with an inclination of 1.8° toward the southwest.

The Quaternary sediments form a lenticular body tapering toward both chain fronts and reaching a thickness >1,000 m below Venice. The sediment core, recovered almost continuously from the VE1 borehole, provided

CORE VE-1

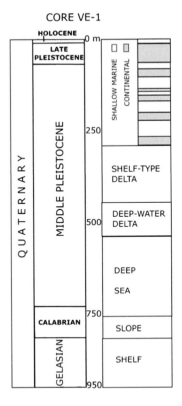

Figure 2.3 Chronostratigraphy and depositional environments of the succession recovered by sediment core VEI.

the paleoenvironmental reconstruction of the last 2 Ma with a good chronological constraint (Figure 2.3).

The depositional area was at first a marine shelf, which subsequently sunk to bathyal depths. From 2 Ma to about 1 Ma, pelagic muds deposited in this basin, and interbedding of sapropels, indicate temporary dysoxic conditions. Above, the placement of turbidite deposits, indicating a massive terrigenous supply from the southern Alps, occurs up to about 0.420 Ma. The basin is gradually filled with deltaic deposits, first distal and then proximal until reaching, in the last 300 m, an alternation of emergence and submersion conditions in relation to the eustatic oscillations (Massari et al. 2004). In this interval, the Venice area was submerged during marine highstand phases and emerged during marine lowstand phases.

Numerous sediment cores carried out for different purposes in the Venice historic center and in the lagoon investigated in detail the sediments of the upper 100 m, from the Upper Pleistocene to the Holocene. The cores allowed recognition and dating of the main evolutionary stages in the

Venetian subsoil, from the last glacial phase to historical times (Tosi et al. 2007a and 2007b).

The detection of the microfaunistic assemblages, combined with the sedimentary facies analysis, allowed definition of the different types of depositional environments. Quantitative analysis of benthic Foraminifera fauna (Albani et al. 2007) was used to recognize, by comparison, the marine and lagoon depositional environments of the late Pleistocene and Holocene ages (Serandrei-Barbero et al. 2006). On these sediments, quantitative microfaunistic analyses similar to those used for the definition of the current marine and lagoon environments were carried out.

In the subsoil near the south lagoon margin, the most recent marine deposits prior to those of the Holocene were found between 64 m and 60 m below the ground surface and present lagoonal facies as shown in Figure 2.3 (Donnici and Serandrei-Barbero 2004). These sediments make up the Correzzola Formation, attributed to the Eemian period, an interglacial period that extended from 127,000 years ago, when the first mixed woodlands were established in Europe, to 115,000 years ago, when open vegetation replaced the forests of Northern Europe and conifers increased significantly toward the south (Kukla et al. 2002). In the subsoil of Venice (Tronchetto island, VE1 borehole), similar marine-lagoonal sediments were found at about 70 m depth (Massari et al. 2004).

Above the Eemian marine deposits, continental deposits referring to the sea-level lowstand during the last glacial phase are present in the Venetian subsoil from 60 m up to the base of the lagoon and coastal sediments of the Holocene age. In this record of continental sediments deposited during the glacial phase, numerous interstadial phases of climatic warming were recognized through pollen analyses (Canali et al. 2007). The culmination of this phase, the LGM (Last Glacial Maximum), occurred from 30,000 to 19,000 years ago (Lambeck et al. 2002). The sedimentary record testifies to the presence of an alluvial plain that occupied the entire area of the northern Adriatic; the paleo-coast was located south of Ancona (Correggiari et al. 1996).

The pollen sequence analyzed in sediment cores below Venice indicates for the last 40,000 years atypical, long glacial period with a predominantly cold dry climate and development of open steppe vegetation with pine (Donnici et al. 2012). This glacial period was interrupted by four phases of climatic warming recorded by the corresponding pollen zones found in the cores, characterized by a significant number of temperate broadleaves and moderate expansion of mountain elements. The average sedimentation rate of the alluvial deposits, approximately 0.9 mm/year during the glacial phase, reached the value of 2.3 mm/year during LGM as shown in Figure 2.4.

On top of Pleistocene alluvial deposits, stratigraphic and sedimentological analyses of numerous sediment cores highlighted the presence of a layer of overconsolidated stiff silty clay. As described in Figure 2.5, this layer, found

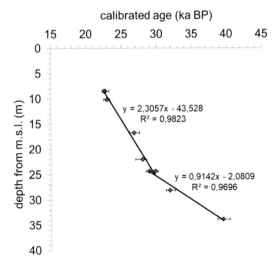

Figure 2.4 Relationship between age and depth in continental deposits from Venice sub-soil (2σ error of calibrated ages is graphically shown). Mean sedimentation rates were calculated in two core/time intervals. A change in the inclination of the curve indicates a change in the sediment accumulation rate around 29 cal ka BP (modified after Donnici et al. 2012).

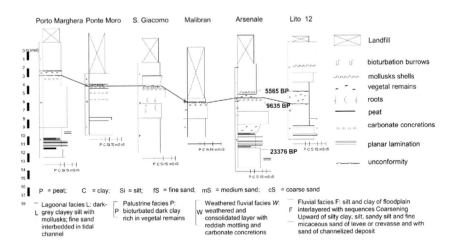

Figure 2.5 Stratigraphy of cores from the lagoon inner margin (Porto Marghera) to the littoral (Lito12). The top of the weathered consolidated layer (*caranto*), highlighted by a solid black line, deepens from the lagoon's inner margin toward the sea (modified after Donnici et al. 2011).

in the subsoil of the Venice lagoon at depths between 3 m and 8 m, shows a maximum thickness of 2–3 m and consists of fine sediments of alluvial origin characterized by the presence of nodules of calcium carbonate and mottling.

This layer is commonly referred to as the *caranto* (from late Latin language: caris = stone). The numerous analyses carried out have allowed researchers to understand that the *caranto* is a paleosol, where the sediments have been altered by subaerial exposure, the traces of oxidation are due to the oscillation of the water table, and the calcareous nodules identify the presence of calcic pedogenetic horizons. Radiometric and pollen analyses, carried out above, below, and inside the *caranto* layer, identified an important stratigraphic gap from about 17,500 to 7,500 years BP, including the Lateglacial and part of the Holocene (Donnici et al. 2011). During this time interval, aggradation did not take place on the Venetian plain and the alluvial sediments remained exposed to atmospheric agents, altering and favoring the formation of soil.

2.3 THE FORMATION OF THE LAGOON

The increase in global temperature and the consequent deglaciation caused the eustatic rise of the sea level with the ascent of the coastline through the Adriatic paleo-plain to a position that corresponds approximately to the current one (Correggiari et al. 1996).

The alluvial sediments deposited during the LGM, during the phase of sea level low stand, were buried by marine deposits during the subsequent marine transgression. During the rapid rise in sea level, barrier-lagoon systems are formed in the northern Adriatic, some of which are partially preserved at different depths on the seabed (Storms et al. 2008; Tosi et al. 2017). The marine transgression reached the Venetian area at different times in the southern, northern, and central zones. Geological surveys indicate that the first marine sediments deposited about 11,000–10,000 years BP under the Chioggia coast (Tosi et al. 2007), about 7,000 years BP in the northern lagoon, where they are found at a depth of 10–12 m (Canali et al. 2007), and 5,500 BP in Venice Arsenale (Donnici et al. 2012).

At the peak of the transgressive phase, the coastline reached a much more internal position than the current one. During the marine highstand, the river supply favored a progradation toward the sea of the coastline, which shifted over time until it reached its present location. Behind the coast that advanced toward the sea, a first, small, lagoon basin was forming, which gradually expanded.

The foraminiferal assemblages of lagoon sediments allowed us to reconstruct the evolution of the lagoon basins and to date their formation. The marine transgression that gave rise to the current lagoon is marked by a shell debris layer indicative of the high-energy transport due to the submersion of the Pleistocene alluvial plain by marine waters. The characteristics of this oldest lagoon phase are those of a closed basin, poorly oxygenated, which subsequently evolves into an open lagoon, evidenced by the greater richness of species indicative of more favorable conditions for life.

About 2,500 years ago, a greater sediment supply led to coastline migration toward the sea. As a result of the presence of lagoon morphologies and the ancient coastline, still found at depths between 2.5 m and 4.8 m above mean sea level, the lagoon turned toward conditions of greater confinement due to the progressive decrease of marine influence.

Later, in addition to the expansion toward the sea, the lagoon basin also extended toward the mainland, due to the lowering of the soil not compensated by fluvial contribution, giving the Lagoon of Venice its current configuration.

The sediments were deposited in the lagoon environment according to the 1‰ inclination of the ancient plain. Therefore, the Holocene succession has a wedge-shaped profile, with thicknesses increasing proceeding from the continental margin toward the sea. The average sedimentation rate in the Venice lagoon varies when different time periods are considered, as reported in Figure 2.6.

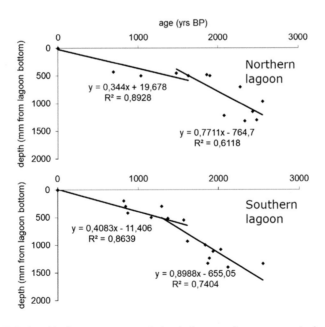

Figure 2.6 Relationship between age and depth from sediment cores in lagoon sediments. Mean sedimentation rate in the northern and southern lagoon in two time intervals: 2,500–1,500 years BP; 1,500 years BP to present (modified from Serendrei-Berbero et al. 2006).

In the southern lagoon, the sedimentation rate is 0.9 mm/year between 2,500 and 1,500 years BP and 0.4 mm/year between 1,500 years BP and the present, while in the northern lagoon, it is 0.8 mm/year and 0.3 mm/year in the time intervals 2,500–1,500 years BP and 1,500 years BP to present, respectively (Serandrei-Barbero et al. 2006).

The drastic decrease in the sedimentation rate in more recent historical times is attributable to the artificial interventions of the Venetians who carried out a gradual shift, starting from the 14th century, of the river mouths.

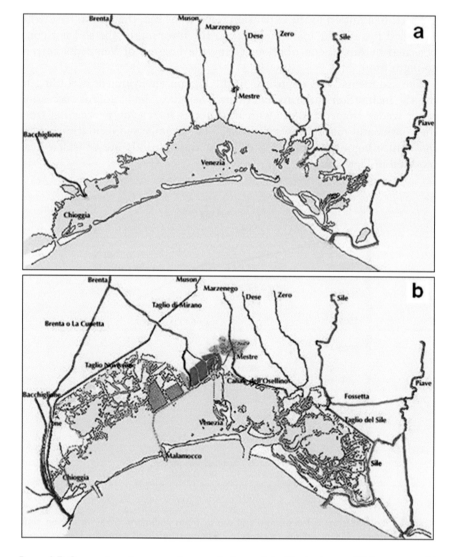

Figure 2.7 Comparison between the morphology of the Venice lagoon 700 years ago (a) and at present (b).

The Brenta and Bacchiglione rivers were diverted away from the Venice lagoon starting from the 16th century. In the 17th century, also the Dese and Sile rivers were diverted from the lagoon and the mouth of the Piave river was shifted northward (Favero et al. 1988). The shape of the three lagoon inlets has been also continuously modified: after the Second World War, the sea bottom at the Malamocco inlet was very deeply dredged to allow the large oil tankers to reach the Venice/Marghera port through the so-called oil canal. Figure 2.7a and b show a comparison between the lagoon morphology in the 12th century with that of the 20th century.

However, the vertical accumulation rate of sediments was also controlled by their location in the lagoon basin and by the bottom morphology, with varying average values in the different lagoon depositional environments. It ranges from 0.3 mm/year in marginal palustrine environments to 2.2 mm/year in tidal channels, settling on the value of 0.6 mm/year in mudflat environments (Donnici et al. 2017).

2.4 THE EVOLUTION OF THE LAGOON AND ITS MORPHOLOGIES

Since its formation, this first lagoon, smaller than the current lagoon, has grown due to the migration of the coasts (fed by the accumulation along the coast of river sediments spilled into the sea), as well as due to the advance of the lagoon itself on its inner edge due to the relative sea level rise. In its expansion, the lagoon has incorporated various morphologies over time, often brought to light and described.

The morphologies buried in the lagoon silts testify to the evolution of the lagoon from its formation to today and, together, preserve evidence of human interventions over time. At various depths, salt marshes, paleochannels, and coastlines have often been highlighted.

Buried littoral, buried salt marshes, and buried channels are some of the different morphologies that can be distinguished.

2.4.1 Buried littoral

In the thickness of the lagoon sediments, the marine sediments are easily identifiable by the presence of shells of marine Foraminifera together with shells of lagoon organisms, as is also typical of the current coastal mixing waters (Albani et al. 1998), and the presence of these shells allow the position of ancient coastlines to be determined.

South of the Brenta river, the oldest coastline, highlighted by the still well-raised coastal dunes, has never exceeded the current internal margin of the lagoon and quickly migrated eastward due to active river transport by sea currents. These coastal dunes are more than 3,000 years old, and north of the Brenta river, this ancient coastline has been located within the southern basin of the lagoon where, between 4.50 m and 7 m below ground surface, there are coastal sands under the sediments of lagoon environment (Favero et al. 1988). In the northern lagoon, northeast of the island of San Erasmo, coastal sand is present below the depth of 2.5 m and the coastline would never have exceeded this limit.

2.4.2 Buried salt marshes

The slow rise of the sea, which was responsible for the formation of the lagoon between 11,000 and 6,000 years ago, was accompanied by the continuous formation of salt marshes (whose characteristics are described in Chapter 7); their growth due to the accumulation of sediments by rivers or tide was insufficient to compensate for the sea level rise, leading to their progressive submersion. In the Venice lagoon, the average sedimentation rate in the last 5,000 years has been 1.4 mm/year, but in the last 1,500 years, it has been only 0.5 mm/year (Canali et al. 2007). Given the recent scarcity of sedimentary inputs, in the last century, there was an altimetric loss of the central lagoon territory, partially due to an anthropic subsidence phenomenon characterized by a maximum settling of 26 cm at the ground surface (Trincardi et al. 2016).

This overall lowering of the bottom caused an area reduction of the morphologies included within the tide range. As reconstructed from historical sources, the salt marshes occupied 149 km² in 1912 (Cucchini 1928) and about 68 km² in 1927. Between 1930 and 1970, their surface was reduced to 47.5 km² while the average depth of the lagoon varied from 0.5 m to 0.6 m (Rusconi 1987). Thus, as calculated by Sarretta et al. (2010), between 1970 and 2002 their overall surface decreased to 32 km², and the average depth (modal depth) dropped to 0.88 m. Despite the general tendency of the lagoon surface covered by salt marshes, there are some examples of newly formed ones.

As further described in Chapter 7, present-day salt marshes are therefore threatened by the rise in sea level and the low sedimentary supply of rivers; however, in the many cores carried out in the historic center of Venice and in the lagoon islands, salt marshes have often been found at the base of human settlements. Buried salt marshes are easily identifiable through the presence of the shells of some microscopic organisms (phylum Foraminifera) exclusive to the intertidal zone, and although they are ephemeral morphologies, they constitute a fixed point in paleoenvironmental reconstructions.

Salt marshes originate from one of the following cases: along the continental margin of lagoons, on soil soaked in salty or brackish water where vegetation has settled; along lagoon canals, where the tidal currents, overflowing, deposit the suspended material; by marine ingression on freshwater marshes.

2.4.3 Buried channels

Many traces of tidal channels are preserved in the thickness of lagoon sediments. Tidal channels play a key role in lagoons and estuaries as they are responsible for the exchange of water with the open sea. The lagoon hydrographic network is closely linked to the morphological characteristics of the basin, such as its width and depth or the number and depth of the inlets. Therefore, variations in position, orientation, shape and size, and the number of buried channels reflect the morphological changes that took place in the lagoon during its evolution.

Geophysical prospecting and sediment cores highlight a dense network of buried channels crossing the Venice lagoon area. Some of them, reconstructed with great detail, have provided valuable information on their morphology, evolution mode, and lifetime (Mandricardo et al. 2007; Mandricardo and Donnici 2014; Bellizia et al. 2021).

A large, buried meander, which has left no visible trace, has been identified just south of Venice (Donnici et al. 2017). The most recent path of the meander has been documented in historical maps, while its oldest path has been identified and reconstructed by geological surveys.

The dating of the innermost side of the eastern part of the meander, at –5.9 m below m.s.l., suggests that this channel was already existing about 4,200 years BP, and from historical maps, the meander was active until 1901.

The comparison of the buried channels with the current ones revealed that the current number of active channels is lower than that of the past, and in some cases, their morphology has been simplified by artificial interventions.

2.4.4 The buried traces of human presence

The lagoon deposits also contain traces of anthropic interventions such as soil reinforcements, foundations, etc., linked to the progressive urbanization of the city.

The lagoon's natural sediments emerge at its continental edge and are about 10 m thick near the littoral. These sediments contain numerous testimonies of human presence in the area: from the flint arrowhead with double-sided processing, found east of the island of Lazzaretto Nuovo and dated to the second millennium BC; the many testimonies dating to the Roman age found extensively in the lagoon area; interventions related to urbanization, such as the fill material used ubiquitously to raise the height of pavements, or the willow branches used to bind canal banks, dating to the 13th century; up to the foundations of the existing buildings (Serandrei-Barbero et al. 2001).

During the 12th and 13th centuries, wood, used as a building material in the early stages of urbanization, was gradually replaced by stone materials and bricks, while at the same time assuming a fundamental role in the laying of foundations. Wooden foundations are addressed in further detail in

Chapter 8. In the Venetian subsoil, the depth of the *caranto* layer, linked to the slope of the paleoplain and to the preexisting morphologies, is between 4 m and 8 m (Donnici et al. 2011). This depth is rarely reached by the foundations and only in monumental buildings.

2.5 MINERALOGY OF SEDIMENTS

With reference to the mineralogical composition of the local sediments, in the past several mineralogical investigations have been carried out on samples collected from different sites within the lagoon, such as from the sites of Tronchetto and Motte di Volpego (located in the central lagoon next to the mainland) and Malamocco and Treporti (along the coastline). Some details may be found in Simonini et al. (2006).

In addition, within the investigations necessary to design the MoSE barriers (explained in detail in Chapters 4–6), a thorough mineralogical study was performed on the cores drawn up from a borehole driven at the Malamocco inlet, located in the buried littoral area (Curzi 1995). More particularly, three mineralogical analyses were performed: on the bulk samples, on the sand fraction, and on the clay fraction. The latter was separated through sedimentation in water and then treated with HCl.

Figure 2.8 depicts the profile of the mineralogical composition obtained from the three analyses. Carbonates, mainly a mixture of detrital calcite and dolomite crystals, are generally the most abundant component. Quartz, feldspar, muscovite (2M micas), and chlorite are other significant components.

Sandy soils show two distinct types of petrographic sources, namely the *granitic* or *Padana* province and the *limestone–dolomite* or *Veneta* province. The granitic component, characterized by a siliceous-clastic composition, derives from sediments found in the basins of the Po and Adige rivers. The limestone–dolomite component, which shows prevalently carbonate sediments (dolomite more abundant than calcite), originates from the basins of the Brenta, Piave, Livenza, and Tagliamento rivers.

The *Padana* and *Veneta* contributions are amalgamated, the first prevailing at higher depths (Favero et al. 1973). Sands appear immature, with a mean rounding index, evaluated in accordance with Powers' scale (Powers 1953), ranging between 0.23 and 0.34.

Typical grain shapes of sand and silt are shown in Figure 2.9a and b; quartz and feldspar grains are generally jagged, suggesting they were not transported for any extended length of time. Similar conclusions may be drawn from the fact that in general feldspars and micas are relatively fresh with extensive alteration appearing occasionally on the crystal surface.

Along with the variation of the grain-size distribution from sands to clays, the carbonate and quartz–feldspar fractions decrease when the content in clay minerals increases.

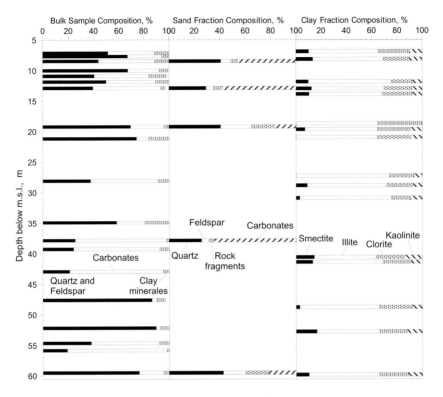

Figure 2.8 Mineralogical composition for bulk samples, sand fraction, and clay fraction of the sediments from the MTS (Simonini et al. 2007).

Figure 2.9 Electron microscope photos for granular soils from Malamocco cores: (a) sand; (b) silt (Cola and Simonini 2002).

As shown in the third column of Figure 2.8, the composition of the carbonate-free clay fraction seems to be rather homogenous at various levels: clay minerals, not exceeding 20% of total weight, are mainly composed of illite (50–60%) with chlorite, kaolinite, and smectites as secondary minerals. X-ray diffractometer traces indicated that small amounts of quartz and feldspar are also present in the clay fraction.

Illite, kaolinite, and chlorite in general appear highly crystallized, suggesting a detrital origin. Small amounts of smectites and interstratified illite–smectite minerals, due to chemical alteration that occurred in the depositional environment, are occasionally present.

Figure 2.10a and b depict the electron microscope photos (Cola 1994) taken on undisturbed silty clay samples collected in Fusina at shallow depths, namely at 4.4 m and 12.3 m below ground level, respectively. The upper Holocenic sample is characterized by particle packets connected in a relatively continuous and regular assemblage with an alveolar structure, probably due to a low-energy lagoon environment. In the deeper sample, the particles seem to be aggregated with an irregular structure in small, well-defined packets and arranged by edge-to-face or edge-to-edge contacts, characteristic of a higher energy depositional environment. This latter feature is in accordance with the findings made by Bonatti (1968) who noted, in samples collected at Motte di Volpego (10 m below ground level), chaotic structures and micro-cross stratification, both characteristic features of sediments deposited from turbulent, rapidly moving masses of water such as rivers.

Figure 2.11a compares the $CaCO_3$ profiles determined at the sites of Motte di Volpego, Tronchetto, Malamocco, and Treporti (Simonini et al. 2006). It can be noticed that the carbonates such as calcite and dolomite are the most prevalent components, with percentages exceeding 50–70% with the exception of about 3–5 m thick layer with reduced $CaCO_3$ concentration, at a depth varying from 20 m to 27 m below mean

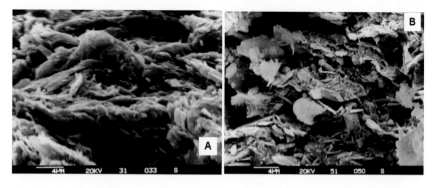

Figure 2.10 Electron microscope photos for cohesive soils from Fusina (Cola 1994): (a) sample at 4.4 m of depth; (b) sample at 12.3 m of depth.

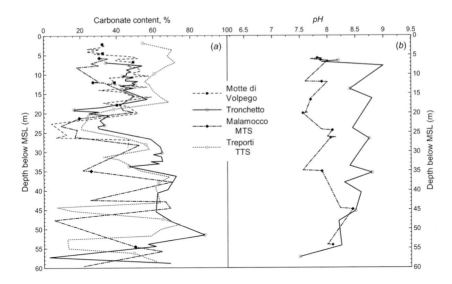

Figure 2.11 (a) Carbonate profiles at Motte di Volpego, Tronchetto, Malamocco, and Treporti; (b) pH profiles at Tronchetto and Malamocco (Simonini et al. 2007).

sea level. Some other low-carbonate episodes exist at both higher and smaller depths, especially at Malamocco test site (MTS). The reductions of $CaCO_3$ together with the presence of peat or organic soils – relatively common in the lagoon area – could be attributed to lacustrine sedimentation episodes.

Profiles of pH measured at the sites of Tronchetto and Malamocco are depicted in Figure 2.10b. Slightly lower values were recorded at Malamocco, especially at intermediate depths. Oscillation of pH at both sites can be noted, with the higher values possibly related to low $CaCO_3$ content, since the presence of the carbonatic environment can neutralize water acidity.

Considering the lagoon formation process, which was characterized by alternating continental and marine depositions, the geological and morphological history of the area, as well as the composition of the sediments, both the complexity and heterogeneity of soil profiles in all the Venetian area become apparent. These qualities have made a more specific geotechnical characterization necessary, including investigations carried out at a special test site, described in Chapter 4.

REFERENCES

Albani, A., Serandrei-Barbero, R. and Donnici, S. 2007. Foraminifera as ecological indicators in the Lagoon of Venice, Italy. *Ecological Indicators* 7(2), 239–253. https://doi.org/10.1016/j.ecolind.2006.01.003.

Albani, A. D., Favero, V. M. and Serandrei-Barbero, R. 1998. Distribution of sediment and benthic foraminifera in the gulf of Venice, Italy. *Estuarine, Coastal and Shelf Science* 46(2), 251–265.

Bellizia, E., Donnici, S., Madricardo, F., Finotello, A., D'Alpaos, A. and Ghinassi, M. 2021. Ontogeny of a subtidal point bar in the microtidal Venice Lagoon (Italy) revealed by three-dimensional architectural analyses. *Sedimentology* 69(3), 1399–1423. https://doi.org/10.1111/sed.12956.

Bonatti, E. 1968. Late-Pleistocene and Postglacial stratigraphy of a sediment core from the Lagoon of Venice (Italy). *Mem. Biog. Adr., Venezia* VII(Suppl.), 9–26.

Canali, G., Capraro, L., Donnici, S., Rizzetto, F., Serandrei-Barbero, R. and Tosi, L. 2007. Vegetational and environmental changes in the eastern venetian coastal plain (Northern Italy) over the past 80,000 years. *Palaeogeography, Palaeoclimatology, Palaeoecology* 253(3–4), 300–316. https://doi.org/10.1016/j.palaeo.2007.06.003.

Carminati, E., Doglioni, C. and Scrocca, D. 2003. Apennines subduction-related subsidence of Venice (Italy). *Geophysics Research Letters* 30(13), 1–4. https://doi.org/10.1029/2003GL017001.

Cola, S. 1994. Caratterizzazione geotecnica delle argille di Fusina. PhD Thesis, University of Padova. (*in Italian*).

Cola, S. and Simonini, P. 2002. Mechanical behaviour of silty soils of the Venice lagoon as a function of their grading properties. *Canadian Geotechnical Journal* 39(4), 879–893.

Correggiari, A., Roveri, M. and Trincardi, F. 1996. Late Pleistocene and Holocene evolution of the North Adriatic Sea. *Quaternary* 9(2), 697–704.

Cucchini, E. 1928. Le acque dolci che si versano nella Laguna di Venezia. (*in Italian*).

Curzi, P. V. 1995. Sedimentological-environmental study of the Malamocco inlet: Final report. *Consorzio Venezia Nuova*, Venezia (*in Italian*).

Donnici, S., Madricardo, F., and Serandrei-Barbero, R. 2017. Sedimentation rate and lateral migration of tidal channels in the Lagoon of Venice (Northern Italy). *Estuarine, Coastal and Shelf Science* 198, 354–366. https://doi.org/10.1016/j.ecss.2017.02.016.

Donnici, S. and Serandrei-Barbero, R. 2004. Paleogeografia e cronologia dei sedimenti tardo-pleistocenici ed olocenici presenti nel sottosuolo di Valle Averto (Laguna Di Venezia, Bacino Centrale). *Lavori della Società Veneta di Scienze Naturali* 29, 101–108.

Donnici, S., Serandrei-Barbero, R., Bini, C., Bonardi, M. and Lezziero, A. 2011. The caranto paleosol and its role in the early urbanization of Venice. *Geoarchaeology* 26(4), 514–543. https://doi.org/10.1002/gea.20361.

Donnici, S., Serandrei-Barbero, R. and Canali, G. 2012. Evidence of climatic changes in the Venetian coastal plain (Northern Italy) during the last 40,000 years. *Sedimentary Geology* 281, 139–150. https://doi.org/10.1016/j.sedgeo.2012.09.003.

Favero, V., Alberotanza, L. and Serandrei Barbero, R. 1973. Aspetti paleoecologici, sedimentologici e geochimici dei sedimenti attraversati dal pozzo VE1bis. *CNR, Laboratorio per lo studio della dinamica delle grandi masse*, Technical report 63, Venezia. (*in Italian*).

Favero, V., Parolini, R. and Scattolin, M. 1988. Morfologia storica della Laguna di Venezia, Arsenale. *Lavori della Società Veneta di Scienze Naturali*, Venezia.

Kukla, G. J., Bender, M. L., Beaulieu, J.-L. de, Bond, G., Broecker, W. S., Cleveringa, P., Gavin, J. E., Herbert, T. D., Imbrie, J., Jouzel, J., et al. 2002. Last interglacial climates. *Quaternary Research* 58(1), 2–13.

Lambeck, K., Yokoyama, Y. and Purcell, T. 2002. Into and out of the last glacial maximum: Sea-level change during oxygen isotope stages 3 and 2. *Quaternary Science Reviews* 21, 343–360.

Madricardo, F. and Donnici, S. 2014. Mapping past and recent landscape modifications in the Lagoon of Venice through geophysical surveys and historical maps. *Anthropocene* 6, 86–96. https://doi.org/10.1016/j.ancene.2014.11.001.

Madricardo, F., Donnici, S., Lezziero, A., De Carli, F., Buogo, S., Calicchia, P. and Boccardi, E. 2007. Palaeoenvironment reconstruction in the Lagoon of Venice through wide-area acoustic surveys and core sampling. *Estuarine, Coastal and Shelf Science* 75(1–2), 205–213. https://doi.org/10.1016/j.ecss.2007.02.031.

Massari, F., Rio, D., Serandrei-Barbero, R., Asioli, A., Capraro, L., Fornaciari, E. and Vergerio, P. P. 2004. The environment of Venice area in the past two million years. *Palaeogeography, Palaeoclimatology, Palaeoecology* 202(3–4), 273–308. https://doi.org/10.1016/S0031-0182(03)00640-0.

Powers, M. C. 1953. A new roundness scale for sedimentary particles. *Journal of Sedimentary Petrology* 23, 117–119.

Rusconi, A. 1987. Variazioni delle superfici componenti il bacino lagunare. *Ufficio Idrografico del Magistrato alle Acque* 160, 140.

Sarretta, A., Pillon, S., Molinaroli, E., Guerzoni, S. and Fontolan, G. 2010. Sediment budget in the Lagoon of Venice, Italy. *Continental Shelf Research* 308, 934–949. https://doi.org/10.1016/j.csr.2009.07.002.

Serandrei-Barbero, R., Albani, A., Donnici, S. and Rizzetto, F. 2006. Past and recent sedimentation rates in the Lagoon of Venice (Northern Italy). *Estuarine, Coastal and Shelf Science* 69(1–2), 255–269. https://doi.org/10.1016/j.ecss.2006.04.018.

Serandrei-Barbero, R., Lezziero, A., Albani, A. and Zoppi, U. 2001. Depositi tardo-pleistocenici ed olocenici nel sottosuolo veneziano: Paleoambienti e cronologia. *Il Quaternario* 14, 9–22. (*in Italian*).

Simonini, P., Ricceri, G. and Cola, S. 2006. Geotechnical characterization and properties of Venice lagoon heterogeneous silts. Invited lecture. Characterization and Engineering Properties of Natural Soils. 29 November–1 December 2006, Singapore, published as preprint in http://hdl.handle.net/11577/2443280.

Storms, J. E. A., Weltje, G. J., Terra, G. J., Cattaneo, A. and Trincardi, F. 2008. Coastal dynamics under conditions of rapid sea-level rise: Late Pleistocene to Early Holocene evolution of barrier-lagoon systems on the Northern Adriatic Shelf (Italy). *Quaternary Science Reviews* 27(11–12), 1107–1123. https://doi.org/10.1016/j.quascirev.2008.02.009.

Tosi, L., Rizzetto, F., Bonardi, M., Donnici, S., Serandrei-Barbero, R. and Toffoletto, F. 2007a. Note Illustrative Della Carta Geologica d'Italia alla scala 1:50.000. Foglio 128 Venezia. [a cura di] Agenzia per la Protezione dell'Ambiente e per i servizi Tecnici APAT, Dipartimento Difesa del Suolo – Servizio Geologico d'Italia.

Tosi, L., Rizzetto, F., Bonardi, M., Donnici, S., Serandrei-Barbero, R. and Toffoletto, F. 2007b. Note Illustrative Della Carta Geologica d'Italia alla scala 1:50.000. Foglio 148–149, Chioggia-Malamocco. [a cura di] Agenzia per la Protezione dell'Ambiente e per i servizi Tecnici APAT, Dipartimento Difesa del Suolo – Servizio Geologico d'Italia.

Tosi, L., Zecchin, M., Franchi, F., Bergamasco, A., Da Lio, C., Baradello, L., Mazzoli, C., Montagna, P., Taviani, M., Tagliapietra, D., et al. 2017. Paleochannel and beach-bar palimpsest topography as initial substrate for coralligenous buildups offshore Venice, Italy. *Scientific Reports* 7, 1. https://doi.org/10.1038/s41598-017-01483-z.

Trincardi, F., Barbanti, A., Bastianini, M., Benetazzo, A., Cavaleri, L., Chiggiato, J., Papa, A., Pomaro, A., Sclavo, M., Tosi, L., et al. 2016. The 1966 flooding of Venice: What time taught us for the future. *Oceanography* 29(4), 178–186. https://doi.org/10.5670/oceanog.2016.87.

Chapter 3

Subsidence of Venice

3.1 INTRODUCTION

Land subsidence is the lowering of the ground surface due to natural and man-induced processes occurring in the shallow and deep ground.

Natural subsidence is controlled by factors acting at different depths, the main ones being the tectonic components, the compaction of the deep ground due to the overlying sediments as well as the consolidation of the newly deposited fresh sediments, especially in coastal areas. Natural subsidence is characterized by slow movements, generally less than 3–5 mm/year, whereas anthropic subsidence may be induced by anthropogenic activities such as hydrocarbon extraction, geothermal exploitation, and pumping from artesian aquifers.

The subsidence in the Venice coastal area is a combination of both the above processes. The anthropic subsidence was caused by water exploitation for industrial purposes, whereas the natural subsidence is a combination of secondary compression, prevalently occurring in the upper soil deposits, and processes of oxidation of surface peaty soils, which particularly affect the marshes and peatlands subjected to tidal cycles.

The entire Northern Italian Adriatic coast, approximately from the Tagliamento River to the city of Rimini, is subsiding at different rates. This subsidence has been induced by the exploitation of water from deep aquifers and by natural processes, but to date not by hydrocarbon extractions. Figure 3.1 shows the rate of subsidence monitored between 1992 and 2000 (Bitelli et al. 2010) along the northern Adriatic coast.

In coastal areas, the rise in sea level, namely the increase in the volume of water in the seas, a process known as eustacy, also contributes to the elevational gap between the land and mean sea level, and this is particularly the case of the city of Venice.

The evolution of this gap in Venice over the last century is shown in Figure 3.2. It should be noted that the actual loss in ground elevation with respect to the mean sea level, including both the lowering of the ground surface due to land subsidence and the increasing of the sea level because

DOI: 10.1201/9781003195313-3

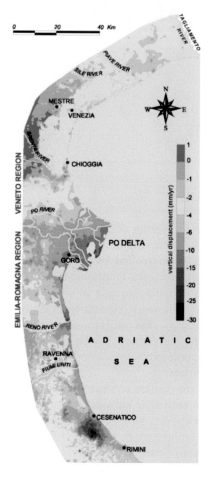

Figure 3.1 Map of ground vertical displacement (mm/year) in the North Adriatic coast obtained with interferometric analysis on ERS-1/2 images for the period 1992–2000 (Bitelli et al. 2010).

of climate change, has reached 3 mm/year over the last decades. From the beginning of the last century, Venice has sunk about 26 cm with respect to mean sea level through the combination of land subsidence and global sea level rise.

3.2 ANTHROPIC SUBSIDENCE IN VENICE

The practice of extracting water from the shallower aquifers underlying Venice was customarily used until 1890, when the domestic water supply was switched to wells on the mainland some 25 km to the northwest.

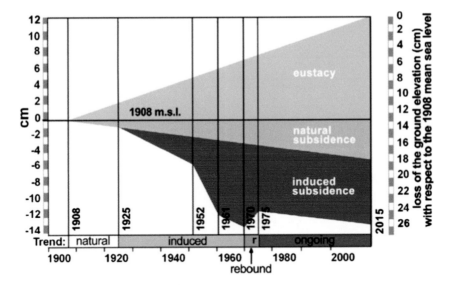

Figure 3.2 Subsidence and eustacy in Venice in the 20th century: natural subsidence in yellow, anthropogenic in red, sea level rise in blue. Subsidence and eustacy combine to a total loss of about 26 cm (Trincardi et al. 2016).

However, as the Mestre industrial complex on the closest mainland developed in the central decades of the last century, large-scale water extraction from the deeper aquifers underlying Venice, up to 350 m deep, ensued from 1938 to 1972.

In these aquifers the piezometric levels were reduced by around 12 m, generating ground surface settlements of up to 150 mm throughout Venice over the same period.

Figure 3.3 shows the settlement rate over the period between 1961 and 1969 also including natural subsidence. The subsidence monitoring began in the 1960s and was initially carried out using benchmarks and classical leveling systems.

The industrial water supply was switched to an aqueduct built for this purpose in 1973. Thereafter, aquifer depletion halted together with the associated ground subsidence, as illustrated in Figure 3.3. Aquifer recharge took about two years, as did a settlement recovery at the ground level of around 20 mm.

Several investigations have been carried out to measure reductions in piezometric levels over time and depth, together with mean surface settlements (Serandrei-Barbero 1972; Carbognin and Gatto 1984; Carbognin et al. 1995; Carbognin and Tosi 2002).

In addition to the VE1 60–950 m depth below m.s.l., two other boreholes, namely VE1-bis, 0–120 m depth; and then VE2 0–400 m depth,

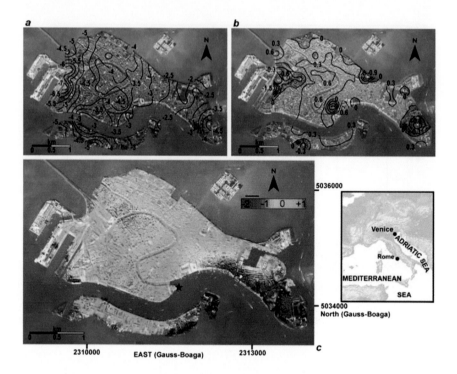

Figure 3.3 Rates of ground vertical displacement (in mm/year) in Venice superimposed on an aerial photograph of the city: (a) from 1961 to 1969 (redrawn after Bergamasco et al. [1993]), (b) from 1973 to 1993 (redrawn after Carbognin et al. [1995]), and (c) from 1992 to 1996. The first two maps are obtained by interpolating scattered values provided by differential leveling; the location of benchmarks is indicated by yellow dots. The contour interval is 0.5 and 0.3 mm/year in (a) and (b), respectively. The 1992–1996 raster map is generated by SRI, with a pixel resolution of 25 m. Subsiding areas are colored from violet to red and stable areas from yellow to blue. Vertical movements of ±0.5 mm/year are in the range of accuracy of the measuring techniques. The black star in (c) shows the position of the Venice tide gauge (Tosi et al. 2002).

were commissioned to classify the soils and to measure mechanical soil properties by means of oedometer tests on undisturbed samples (Ricceri and Butterfield 1974; Rowe 1975).

All these studies provided suitable piezometric level and surface settlement data, as well as relatively reliable information on a representative soil column.

More particularly, by examining continuous cores from the boreholes VE1, VE1-bis, and the upper 100 m of the borehole VE2, it was concluded that within the upper 350 m, the soils are essentially sand (69%), silts (29%), and silty clays (2%). The small quantities of clay and peat present are not considered to contribute significantly to the overall compressibility of the soil column.

Based on the outcomes from these boreholes, settlement back-calculation was carried out independently by Ricceri and Butterfield (1974) and by Rowe (1975).

Since the ratio of the extent of the loaded area to its depth is very large, it may be assumed that settlement predictions in one-dimensional compression, using the piezometric level changes and soil oedometric compressibility parameter, should be relatively simple and should likely agree with the measured settlements.

However, through the use of the VE1/VE1-bis data and oedometer tests on 76 mm rubber-sleeve samples, Ricceri and Butterfield (1974) calculated that in order to reconcile the measured settlements with the known increase in vertical effective stress, the estimated classical mean compression index had to be halved. A suggested explanation of this discrepancy was that either the soil could be returning to virgin compression from an unloading/reloading excursion or that the laboratory-determined compressibility parameters were perhaps too high due to sample disturbance during recovery.

To check the latter point, borehole VE2 was commissioned, in which 60 mm diameter piston sampling to 100 m depth recovered 14 samples for consolidation in the 250 mm 'Rowe's cell'.

The resulting compression curves were very similar to those determined on 76 mm samples at the University of Padova (Ricceri and Previatello 1972). Using parameters from these tests, Rowe (1975) attempted to carry out a settlement prediction. Rowe's data, shown in Table 3.1 (column 7), suggested that the groundwater level drawdown between 1938 and 1970 should have caused 205 mm of settlement. He quotes 133 mm as the measured value over this period even though, from a number of independent records, Ricceri and Butterfield (1974) estimated the value to have been between 140 mm and 145 mm. Rowe's estimate was not particularly suitable with respect to the one proposed by Ricceri and Butterfield (1974).

On the basis of previous studies on soil compressibility (Butterfield and Baligh 1996), Butterfield et al. (2003) have recently proposed a new method. This method is based on the interpretation of the oedometric compression curves in a more suitable $\log e - \log \sigma'_v$ plane, which properly allows effective soil compressibility as well as stress history to be considered. More details can be found in Butterfield et al. (2003). Table 3.1 reports the results of Rowe's calculation (column 7) compared to that (column 9) provided by the new method suggested by Butterfield et al. (2003). The selected one-dimensional compressibility parameter is $m_v = -1/(1 + e_0) \cdot de/d\sigma'_v$, where e = the current void ratio, e_0 = in situ void ratio, σ'_v = vertical effective stress (in Table 3.1, $p'_0 = \sigma'_{v0}$ = geostatic vertical overburden stress). Note the excellent agreement of the new method (153.2 mm) with the measured value (around 150 mm) compared to that of Rowe (205.1 mm).

Table 3.1 Data and settlement predictions

Depth (m)	Mean p'_0 (MN/m²)	Soil type	Thickness (m)	m_v (m²/MN) $\times 10^{-2}$	Drawdown (m)	Settlement (mm)	m_v (m²/MN) $\times 10^{-2}$	Settlement (mm)
25–62	0.40	Silt sand	15.6	75	0.7	8.2	4.5	4.9
			21.4	0.009		0.1	0.09	0.1
62–112	0.81	Silt sand	1.3	4.7	2.7	1.6	4.5	1.6
			48.7	0.09		1.2	0.09	1.2
112–162	1.27	Si/Cl	4.2,	3.2	6.8	9.1	3.0	8.6
		Silt and	8.0	3.2		17.4	2.9	15.8
			37.8	0.009		2.3	0.09	2.3
162–238	1.86	Silt sand	24.3	2.4	9.55	55.5	2.0	46.4
			51.7	0.09		4.4	0.09	4.4
238–307	2.52	Silt and	29.3	2.0	9.00	52.6	1.4	36.9
			39.7	0.09		3.2	0.09	3.2
307–334	2.97	Silt sand	11.0	1.9	7.5	15.7	1.2	9.9
			16.0	0.09		1.1	0.09	1.1
334–389	3.35	Si/Cl	10.0	1.7	3.5	5.9	1.1	3.9
		Silt sand	45.0	1.7		26.8	1.1	17.3
					Total	205.1		153.2

3.3 NATURAL SUBSIDENCE IN VENICE

As explained in Chapter 2, Venetian soils have undergone continuous gravitational compaction with, more recently, phases of continental sedimentation (42,000–16,000 years BP) and erosion (16,000–6,000 years BP), followed by re-immersion during the Flandrian transgression. During the emersion period, the surface of the deposit was exposed and desiccated. Although the primary consolidation phase of the gravitational compaction process is certainly complete, recent monitoring shows that the mean surface subsidence over the area of the lagoon continues to increase due to a combination of secondary compression (creep) in the soil column, sinkage of the periappenninic trough and tectonic movement in the northern Adriatic basin, and develops over geological time scales (Tosi et al. 2010).

It acts with different rates ranging from 0.6 to 1.6 mm/year, depending on the time span considered (Tosi et al. 2013).

Historical evidence of the human response to natural subsidence, attributable to the creep strain of the soil solid skeleton, is the raising of building floors and street pavements required over centuries to combat flooding. An example is provided by the photo in Figure 3.4 where it is evident that the 16th-century pavement has been elevated with respect to the early medieval column base. Over about 500 years, the pavement was raised by around 50 cm, which indicates an average settlement rate of 1.1 mm/year.

Due to the need to control the ground movements in Venice and its lagoon, the monitoring of land subsidence, especially the natural one, has been significantly improved through the use of space-borne observation techniques based on Global Positioning System (GPS) and synthetic aperture radar (SAR) interferometry.

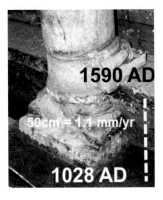

Figure 3.4 Evidence of past subsidence occurrence: rising of the base of an early medieval column (courtesy Laura Carbognin).

3.4 THE SUBSIDENCE-MONITORING NETWORK OF THE VENICE COASTLAND

The subsidence-monitoring network of the Venice coastland increased from a few hundred benchmarks measured by leveling to a few hundred thousand reflectors detected by SAR-based interferometry using ERS and ENVISAT satellites. Later, the use of TerraSAR-X, COSMO-SkyMed, and Sentinel 1 generations provided significant advancements in the knowledge of land subsidence in the Venice area.

The overall picture of the ground movement in the Venice lagoon provided by Figure 3.5 (Tosi et al. 2013) highlights a large variability in terms of land subsidence, with the rates generally ranging from negligible values (less than 1 mm/year) to 6 mm/year but locally exceeding 10 mm/year.

The largely heterogeneous nature of the shallow Holocene deposits is strongly responsible for the high heterogeneity of land movements at the regional scale (Tosi et al. 2016). The highest settlements are localized in correspondence to the littoral and are due to anthropogenic activities. One of the main mechanisms responsible for high sinking rates is the load of newly built-up areas and engineering structures. Inside the lagoon, the higher sinking rates are often linked to the presence of stone embankments bounding fish farms and the reconstruction of artificial salt marshes (Da Lio et al. 2018).

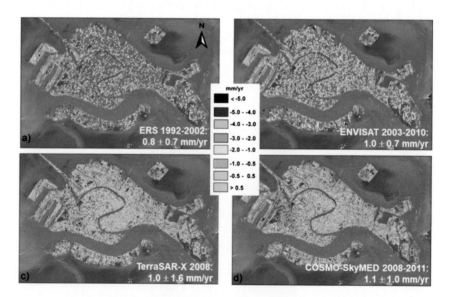

Figure 3.5 Mean displacement rates (mm/year) in Venice. Negative values indicate settlement, positive mean uplift. The average values for each map are provided (Tosi et al. 2013).

Regarding Venice's historical center, it has been revealed that the local variability of the subsidence pattern strictly reflects the urbanization growth that began around 900 AD after the first island of Venice was settled (Tosi et al. 2013). The older parts of the city show lower sinking rates, while areas developed later are highly susceptible to compaction. In fact, the historical center was built on already well-consolidated sandy islands, whereas later expansions occurred through land reclamation, by filling parts of the soft muddy tidal flats and lagoon channels.

Recently, the proper combination of SAR-based interferometry products, retrieved by different sensors (X- and C-band), allowed recognition of natural subsidence and ground movement due to anthropogenic activities. Results revealed the occurrence of short-term (1–2 years) and highly localized (e.g., single palace) sinking, up to 6 mm/year, linked to building renovations and maintenance work in the city, as reported in Figure 3.6 (Tosi et al. 2013; Tosi et al. 2018).

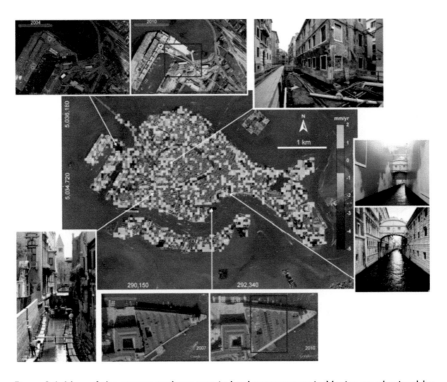

Figure 3.6 Map of the recent anthropogenic land movements in Venice as obtained by subtracting the long-term ground movements from the short-term ground movements. Negative and positive rates indicate areas where human activities are either responsible for or reduce land settlements, respectively. A few examples of human activities that may influence subsidence values are shown in the photographs (available from http://www.insula.it/ and Google Earth images) (Tosi et al. 2018).

Regarding the lagoon basin, the use of SAR interferometry together with in situ measurements allowed a step forward in the characterization of land subsidence of salt marshes. As already introduced in previous chapters, salt marshes are fragile morphologies that represent the result of interactions between climate changes and sedimentary processes in which the presence of halophytic vegetation species and the related production of organic matter play a key role. Corner reflectors (Strozzi et al. 2013), benchmarks monitored by very high-precision leveling, and the surface elevation table apparatus, combined with short-term accretion feldspar marker horizons, revealed, as shown in Figure 3.7, high heterogeneity of the superficial subsoil dynamics with displacements that range from small uplifts to subsidence rates of more than 20 mm/year (Da Lio et al. 2018). As highlighted in Figure 3.7, it has also been pointed out that there is a significant difference between the subsidence in the marshes with and without vegetation (represented in the graphs by differently colored points), revealing that higher sinking rates occur on bare salt marshes.

Regarding the littoral area, the construction of mobile barriers at the three inlets (as part of the MoSE project, described in Chapter 6) raised concerns as to possible important sinkage caused by the load of complementary structures such as jetties, breakwaters, locks, and an artificial island on the Quaternary deposits.

Monitoring of land subsidence at high spatial resolution provided relevant information for the geotechnical characterization of the coast in relation to the stability of large coastal structures (Tosi et al. 2018). An example of ground movements detected at the inlet of Malamocco is shown in Figure 3.8. A back-analysis of the observed settlements of this jetty was discussed by the author of this book and provided in Tonni et al. (2016).

In conclusion, land subsidence is a process that was originally part of the formation of the Venetian lagoon but has gradually become its greatest danger to face. Once the stability of the historical center was verified, the attention to the process moved to other parts of the lagoon, and a picture of highly heterogeneous land subsidence was obtained. Owing to its uniqueness, the city of Venice and its lagoon have been the subject of methodological experiments that use cutting-edge technologies to monitor ground movements and have brought about important progress in the knowledge of the process of subsidence. The results of the monitoring provide not only an up-to-date picture of the magnitude of the process but also the support for an accurate geotechnical characterization of the various lithotypes that make up the subsoil of the Venice lagoon.

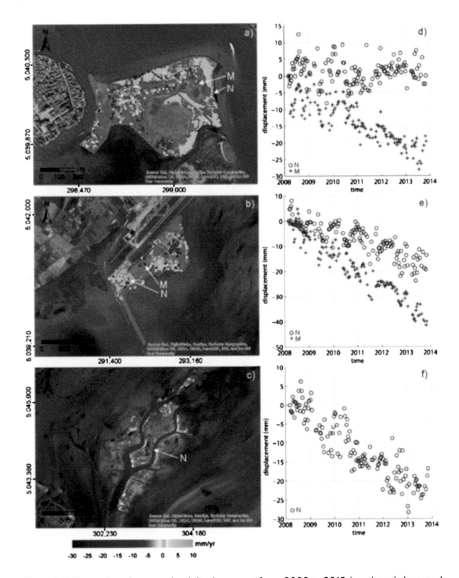

Figure 3.7 Examples of average land displacement from 2008 to 2013 (mm/year) detected by PSI on the salt marshes: (a) Burano; (b) Tessera; and (c) Cenesa. Positive values indicate uplift, while negative values indicate land subsidence. In (a–c) the natural portions of the salt marshes are shaded green, while man-made parts are shaded red. N and M refer to selected PTs on natural and man-made salt marshes, respectively, for which the displacement time series are provided in (d–f). Base map source: Esri, DigitalGlobe, GeoEye, Earthstar Geographics, CNES/Airbus DS, USDA, USGS, AEX, Getmapping, Aerogrid, IGN, IGP, swisstopo, and the GIS User Community (from Tosi et al. 2018).

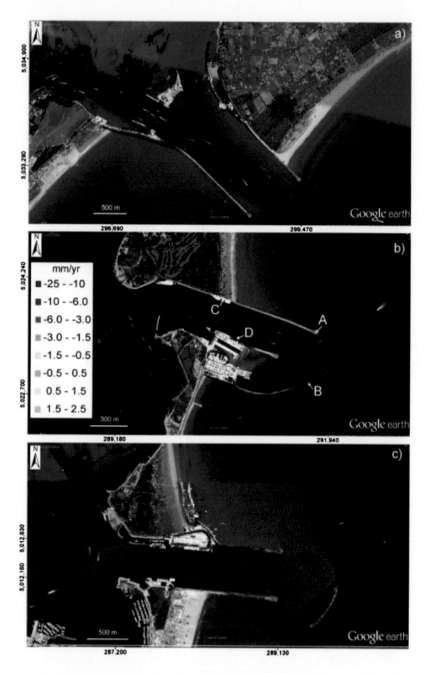

Figure 3.8 Mean displacement rates from TerraSAR-X interferometry: between August
2012 and November 2013 at the Lido inlet (a) and between March 2008 and
November 2013 at the Malamocco (b) and Chioggia (c) inlets. The back-
grounds are Google Earth images acquired in 2012. Movements are in the
satellite line-of-sight direction, negative values indicate settlement, and positive indicate uplift (from Tosi et al. 2018).

REFERENCES

Bitelli, G., Bonsignore, F., Carbognin, L., Ferretti, A., Strozzi, T., Teatini, P., Tosi, L. and Vittuari, L. 2010. Radar interferometry-based mapping of the present land subsidence along the low-lying northern Adriatic coast of Italy. *Land Subsidence, Associated Hazards and the Role of Natural Resources Development, Proceedings of EISOLS*, 339, 279–286.

Butterfield, R. and Baligh, F. 1996. A new evaluation of loading cycles in an oedometer. *Géotechnique* 46(3), 547–553.

Butterfield, R., Gottardi, G., Simonini, P. and Cola, S. 2003. A new interpretation of the compressibility of Venetian silty-clay soils. *ICOF 2003*, Dundee, 2–5 September 2003, Thomas Telford.

Carbognin, L. and Gatto, P. 1984. An overview of the subsidence of Venice. *Proceeding 3rd International Symposium on Land Subsidence, IAHS*, 151, 321–328.

Carbognin, L. and Tosi, L. 2002. Interaction between climate changes, eustacy and land subsidence in the North Adriatic region, Italy. *Marine Ecology-Progress Series* 23, 38–50.

Carbognin, L., Tosi, L. and Teatini, P. 1995. Analysis of actual land subsidence in Venice and its hinterland (Italy). In F. B. J. Barends, F. J. J. Frits and F. H. Schrader (Eds.), *Land Subsidence*. Wallingford: IAHS Publ., 129–137.

Da Lio, C., Teatini, P., Strozzi, T. and Tosi, L. 2018. Understanding land subsidence in salt marshes of the Venice Lagoon from SAR Interferometry and ground-based investigations. *Remote Sensing of Environment* 205, 56–70.

Rowe, P. W. 1975. *Inherent Difficulties in the Application of Geotechnical Science*, Proceeding Symposium on Recent Developments in the Analysis of Soil Behaviour, 3–29, University New South Wales, Australia.

Ricceri, G. and Butterfield, R. 1974. An analysis of compressibility data from a deep borehole in Venice. *Géotechnique* 24(2), 175–192.

Ricceri, G. and Previatello, P. 1972. *Caratteristiche geotechniche del sottosuolo della Laguna Veneta*, Memorie e Studi dell'Istituto di Costruzioni Marittime e del Centro Geotecnico Veneto, No.93, 1972 *(in Italian)*.

Serandrei Barbero, R. *Indagine sullo sfruttamento artesiano nel Comune di Venezia, 1846–1970.* CNR-ISDGM Techn. Rep. 31, Venezia 1972 *(in Italian)*.

Strozzi, T., Teatini, P., Tosi, L., Wegmüller, U. and Werner, C. 2013. Land subsidence of natural transitional environments by satellite radar interferometry on artificial reflectors. *Journal of Geophysical Research: Earth Surface* 118, 1177–1191.

Tonni, L., García Martínez, M. F., Simonini, P. and Gottardi, G. 2016. Piezocone-based prediction of secondary compression settlements of coastal defence structures on natural silt mixtures. *Ocean Engineering* 116, 101–116.

Tosi, L., Carbognin, L., Teatini, P., Strozzi, T. and Wegmüller, U. 2002. Evidence of the present relative land stability of Venice, Italy, from land, sea, and space observations. *Geophysical Research Letters* 29(12): 3–1. https://doi.org/10.1029/2001GL013211.

Tosi, L., Teatini, P. and Strozzi, T. 2013. Natural versus anthropogenic subsidence of Venice. *Scientific Report* 3, 2710.

Tosi, L., Teatini, P., Strozzi, T., Carbognin, L., Brancolini, G. and Rizzetto, F. 2010. Ground surface dynamics in the northern Adriatic coastland over the last two decades. *Rendiconti Lincei Scienze Fisiche e Naturali* 21, 115–129.

Tosi, L., Da Lio, C., Strozzi, T. and Teatini, P. 2016. Combining L- and X-Band SAR interferometry to assess ground displacements in heterogeneous coastal environments: The Po River Delta and Venice Lagoon, Italy. *Remote Sensing* 8, 308.

Tosi, L., Lio, C. D., Teatini, P. and Strozzi, T. 2018. Land subsidence in coastal environments: Knowledge advance in the Venice coastland by TerraSAR-X PSI. *Remote Sensing* 10(8), 1191.

Trincardi, F., Barbanti, A., Bastianini, M., Benetazzo, A., Cavaleri, L., Chiggiato, J., Papa, A., Pomaro, A., Sclavo, M., Tosi, L. and Umgiesser, G. 2016. The 1966 flooding of Venice: What time taught us for the future. *Oceanography* 29(4), 178–186. https://doi.org/10.5670/oceanog.2016.87.

Chapter 4

Site and laboratory geotechnical investigations

4.1 INTRODUCTION

Over the last 50 years, the soil deposits that compose the ground on which the historic city and the entire lagoon lie have been intensively investigated by using standard and advanced site and laboratory geotechnical investigations.

As already introduced in Chapter 2, the first comprehensive investigation was carried out in the early 1970s, to characterize the Venetian soils down to a depth of 1,000 m. This investigation was used to sketch a detailed soil profile and to measure hydro-mechanical properties for modeling the subsidence phenomenon.

Between the 1970s and 1980s, several geotechnical investigations were carried out in the Venetian plain, mainly in order to design the foundations of several new, large industries located on the mainland. This industrial area is situated in the towns of Mestre and Marghera and faces the inner lagoon. However, no consistent framework characterizing soil behavior was proposed at that time.

Only within the huge MoSE project, better described in Chapter 6, did it become possible to study and characterize the Venice lagoon subsoil in a systematic way. This was necessary in order to create a suitable design of the foundations for the tilting gates at the lagoon inlets and to actuate many other interventions aimed at protecting the city and the lagoon from further environmental damage.

These investigations determined that the Venetian deposits in any area of the lagoon are mostly composed of a predominant silt fraction combined with sand and/or clay according to chaotic and erratic patterns, rapidly variable even in nearby sites, thus leading to highly heterogeneous soil conditions and associated scale effect in sketching local soil profiles.

On the basis of the above observations, in the early 1990s, the Venice Water Authority decided to concentrate research efforts and financing at one special test site, located at the Malamocco Inlet (the *Malamocco test site* – MTS, whose location is represented in Figure 4.1), where a series of investigations including boreholes (BH), piezocone (CPTU), seismic piezocone

DOI: 10.1201/9781003195313-4

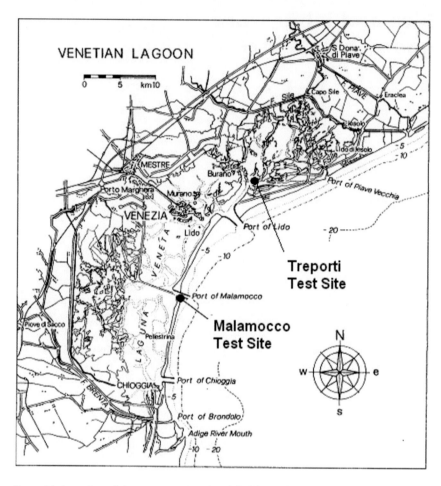

Figure 4.1 Location of the two test sites (modified from Cola and Simonini 2002).

(SCPTU), dilatometer (DMT), self-boring pressuremeter, and cross hole tests (CH) were performed on contiguous verticals within a limited area. To retrieve soil samples while reducing the disturbance effect, a new, large-diameter piston was also expressly designed and used. Nevertheless, due to reasons such as the poorly structured nature of these soils, the effect of sampling stress relief, and a relevant scale effect, it was not entirely possible to achieve a stress history reconstruction, nor a satisfactory stress–strain–time mechanical characterization, necessary for reliable settlement calculations.

In addition, the geotechnical laboratory tests carried out at MTS (Cola and Simonini 2002; Biscontin et al. 2001, 2007) emphasized that due to the high heterogeneity of these materials, a relatively large number of tests were required to define even the simplest properties with a certain degree of accuracy.

In order to solve some of these issues, after some time a new geotechnical investigation project was financed by the Venice Water Authority. This project's aim was to measure directly through a load test the stress–strain–time properties of these heterogeneous soils. The selected test site is located in Treporti (the *Treporti test site* – TTS, also represented in Figure 4.1), a fishing village facing the lagoon close to the Lido Inlet, where the Lido barrier would later be installed (see Chapter 6). At the TTS, a reinforced earth embankment with vertical walls, measuring 40 m in diameter, was built. During and after construction, the ground displacements were measured by using plate extensometers, differential micrometers, GPS, and inclinometers (Simonini et al. 2006; Jamiolkowski et al. 2009). Boreholes with undisturbed sampling, CPTU/SCPTU, and DMT/seismic DMT (SDMT) were used to characterize soil profile and evaluate laboratory geotechnical properties for a comparison with those directly measured in situ.

4.2 THE MALAMOCCO TEST SITE

4.2.1 Soil profile and basic properties

The soil profile and basic material properties, depicted in Figure 4.2, have been determined on the basis of an extensive geotechnical laboratory campaign carried out on both 100 mm and 220 mm diameter samples. Note that the ground surface at MTS lies around 11 m below mean sea level.

The investigation at MTS found that the Venice lagoon soils may be grouped into three classes, that is, medium to fine sand (SP-SM – 35%), silt (ML – 20%), and very silty clay (CL – 40%). The remaining 5% is shared between medium plasticity clay and organic soils (CH, OH, and Pt). In two-thirds of the analyzed samples, silt accounts for over 50% of the total composition.

Sands are relatively uniform; however, finer sediments are more graded, that is, the finer the material, the higher the coefficient of non-uniformity $U = D_{60}/D_{10}$, which increases at decreasing mean particle diameter D_{50}. Liquid limit is $LL = 36 \pm 9\%$ and plasticity index is $PI = 14 \pm 7\%$, whereas the Activity $A = PI/CF$ (CF = clay fraction) is low, with the great majority of samples falling in the range $0.25 < A < 0.50$. Saturated unit weight γ_{sat} ranges between 17 kN/m³ and 21 kN/m³ approximately, whereas void ratio e_o ranges between 0.6 and 1.0; the higher values are due to the presence of thin organic layers.

Preconsolidation (precompression) stress and overconsolidation ratio (OCR) have been evaluated from oedometric tests, carried out on CL-CH specimens. Due to sample disturbance, the estimate of preconsolidation stress was not straightforward, and some CH specimens excluded. The soils are slightly overconsolidated (OC), except the already mentioned highly

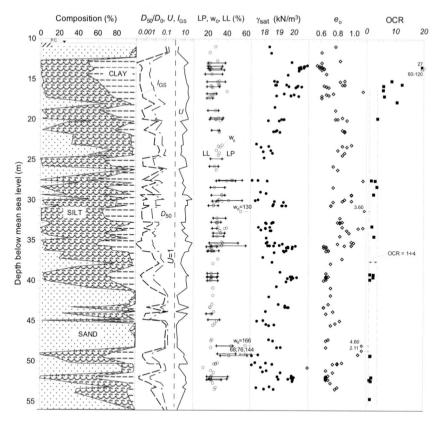

Figure 4.2 Basic soil properties at MTS (modified from Cola and Simonini 2002).

OC formation (*caranto*) that marks the separation between Pleistocene and Holocene sediments.

4.2.2 Stress–strain behavior

Different classes of soils have been tested in the laboratory using oedometric, triaxial, the latter equipped with non-contact displacement sensors and bender elements (BE), and resonant column (RC) apparatuses. The extensive laboratory program is described in Cola and Simonini (2002).

Figure 4.3a depicts some oedometric compression curves for CL, ML, and SP-SM soil classes. CL and ML specimens have been trimmed from undisturbed cores, whereas SM-SP specimens have been prepared using the freezing technique. Note the very gradual transition to the virgin compression regime, even for CL sediments, which makes the estimation of preconsolidation stress problematic. In the case of Venetian soils, the one-dimensional stress–strain response has been evaluated also for sands

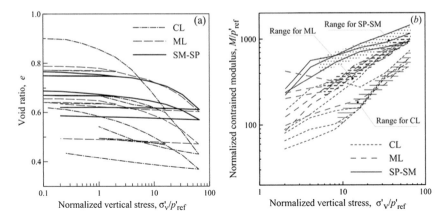

Figure 4.3 (a) Void ratio vs. normalized vertical effective stress (Cola and Simonini 2002); (b) Normalized constrained modulus vs. normalized vertical effective stress (Cola and Simonini 2002).

SP-SM. Note the gradual transition at higher stresses, very similar to that observed for silts.

It is interesting to interpret the oedometric curves in terms of constrained modulus M as a function of vertical effective stress σ'_v (Janbu 1963),

$$\frac{M}{p'_{ref}} = C \cdot \left(\frac{\sigma'_v}{p'_{ref}} \right)^m \tag{4.1}$$

where C and m are two experimental constants and p'_{ref} a reference stress (100 kPa).

Trends of M as a function of vertical stress for the three classes of soil are depicted in Figure 4.3b. The M of CL class, which is less than half that of silts, is characterized by the lowest values, whereas the highest values correspond to the SM-SP class.

Typical results of TX-CK₀U tests for the classes SM-SP and CL are depicted in Figure 4.4a, in terms of deviatoric stress q vs. axial strain ε_a and pore pressure u vs. ε_a, while Figure 4.4b shows the stress paths in the p'-q plane (p' = mean effective stress).

Behavior of CL soils is very similar to that of SM-SM with both contractant and dilatant phases. The phase transformation point (PTP) (Tatsuoka and Ishihara 1974, Ishihara et al. 1975) marks the end of the contractant phase; after that the specimen shows dilatant behavior.

In just a few of the tests performed on CH specimens, the behavior turned out only contractive.

The critical state angles ϕ'_c as well as the angle ϕ'_{PTP} at the point transformation phase for all the tested samples are approximately in the range

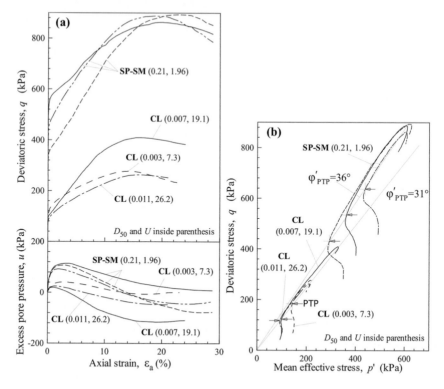

Figure 4.4 Typical undrained triaxial behavior for SP-SM and CL soils: (a) q and u vs. axial strain; (b) p'–q stress paths (modified from Cola and Simonini 2002).

30–34° for CL, 33–37° for ML, and 34–39° for SM-SP, showing some superposition among different soil classes. No noticeable difference has been observed between ϕ '$_{PTP}$ and ϕ '$_c$, with no differences among the three classes of soil and drainage conditions.

Figure 4.5 reports the difference between peak and critical friction angle (ϕ '$_p$ – ϕ '$_c$) in triaxial tests vs. the mean effective stress at failure p'_f (in the figure, CF is the clay fraction).

It is interesting to notice a general decrease (ϕ '$_p$ – ϕ '$_c$) as a function of p'_f for both sandy and silty soils (the shaded area contains most of the data from SM-SP and ML specimens) without any particular distinction between these two classes. Very silty clays CL are characterized by the difference (ϕ '$_p$ – ϕ '$_c$) not depending on p'_f.

The equation proposed by Bolton (1986, 1987) for deriving ϕ '$_p$ in triaxial compression was considered to interpret the triaxial test results:

$$\phi'_P - \phi'_c = 3D_R \left[\left(Q - \ln p'_f \right) - 1 \right] \tag{4.2}$$

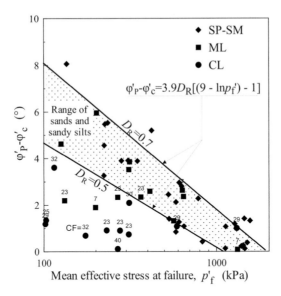

$$\varphi'_P-\varphi'_c=3.9D_R[(9 - \ln p_f) - 1]$$

Figure 4.5 Difference between peak and critical angles vs. mean effective stress at failure in triaxial tests (modified from Cola and Simonini 2002).

where $D_R = (e_{max} - e)/(e_{max} - e_{min})$ is the relative density and Q an experimental constant (Bolton suggested $Q = 10$ for quartz and feldspar sands and 8 for limestone). The equation was adapted to Venice soils as follows:

$$\phi'_p-\phi'_c = 3.9D_R\left[\left(9 - \ln p'_f\right)-1\right] \tag{4.3}$$

The above equation is plotted in Figure 4.5 for $D_R = 0.5$ and $D_R = 0.7$, a typical range of density for Venice sands. The value 3.9, used here instead of 3 suggested by Bolton, may be partially justified by considering that the Venetian sands and silts are composed of angular grains, whereas the data used by Bolton mostly referred to sands characterized by a lower degree of angularity.

Small strain shear stiffness G_{max} has been estimated using resonant column tests (RC), and shear wave velocity measurements are performed in triaxial tests using bender elements (BE). The experimental data were interpreted as a function of p', and e according to the relationship proposed by Hardin and Black (1969) and Hardin and Drnevich (1972) (D is a material constant and n is an exponent).

$$\frac{G_{max}}{p'_{ref}} = D \cdot \frac{\left(2.97 - e\right)^2}{(1 + e)} \cdot \left(\frac{p'}{p'_{ref}}\right)^n \tag{4.4}$$

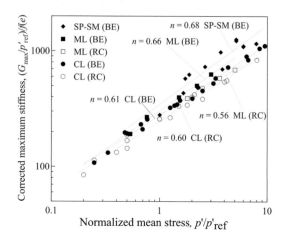

Figure 4.6 Maximum shear stiffness vs. mean effective stress (Cola and Simonini 2002).

neglecting the effect of overconsolidation/precompression, which for sands low-plasticity and slightly OC cohesive soils are very small (Hardin and Drnevich 1972).

Figure 4.6 shows the ratio $G_{max}/(f(e) \cdot p'_{ref})$ as a function p'/p'_{ref}. The exponent $n = 0.60$ could be reasonably assumed as representative for all three classes of soil.

Regarding time-dependent behavior, ML and CL classes are characterized by primary consolidation coefficient c_v, in the range 8×10^{-8} to 4×10^{-6} m^2/s, suggesting that Venetian silts are relatively free draining materials.

Secondary compression coefficient C_α of ML and CL estimated from the one-dimensional compression tests varies between 4×10^{-4} and 4×10^{-3}, that is, in a relatively large range depending on stress state and material density and is more affected by the stress history rather than by grain size composition (Simonini 2004).

In accordance with Mesri and Godlewski (1977) and Mesri et al. (1995), the ratio C_α/C_c (C_c is the local slope of the e-logσ'_v curve) has been evaluated. For Venetian materials, average C_α/C_c is around 0.028 (upper value = 0.050 and lower value = 0.015), with no appreciable differences between the classes CL, ML, and SP-SM soils (Simonini 2004).

More detailed information on the stress–strain response may be found in Cola and Simonini (2002), Simonini (2004), and Simonini et al. (2006).

4.2.3 The grain size index

Figure 4.7 reports the profile of mean particle diameter D_{50}/D_0 ($D_0 = 1$ mm) and non-uniformity coefficient $U = D_{60}/D_{10}$ against depth. It is important to notice that at higher D_{50}, i.e., for sands, the Venetian soil is relatively

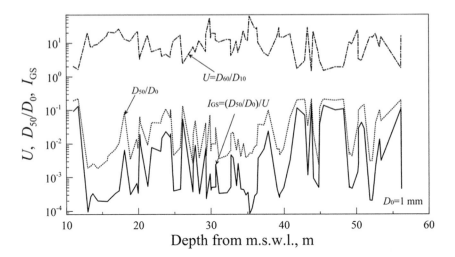

Figure 4.7 Grain size index vs. depth at MTS (Cola and Simonini 2002).

uniform. However, moving toward silts and silty clays, the grain size distribution displays a larger range, that is, at decreasing D_{50}.

The opposite D_{50}-U trends, together with the common mineralogical origin of the Venetian soils (sediments coming from the surrounding Alps originated from siliceous-calcareous sands up to the low-plasticity silty clays), suggested the introduction of a new parameter, referred to as grain size index I_{GS} (Cola and Simonini 2002), to which relevant time-independent mechanical properties can be related.

$$I_{GS} = \frac{D_{50} / D_0}{U} \tag{4.5}$$

The application of I_{GS} was limited to the range $8 \times 10^{-5} \leq I_{GS} \leq 0.12$ and CF $< 25\%$.

The following parameters have been related to I_{GS}:

- C and m of Janbu's Equation (4.1);
- Critical state parameters, ϕ'_c, λ_c, e_{ref};
- D parameter of Hardin and Drnevich's Equation (4.4).

The following relationships have been obtained:

$$C = (270 \pm 30) + 56 \cdot \log I_{GS} \tag{4.6}$$

$$m = (0.30 \pm 0.10) - 0.07 \cdot \log I_{GS} \tag{4.7}$$

$$\phi'_c = (38.0 \pm 2.0) + 1.55 \cdot \log I_{GS} \quad (°) \tag{4.8}$$

$$\lambda_c = (0.152 \pm 0.04) - 0.037 \cdot \log I_{GS} \tag{4.9}$$

$$e_{ref} = (1.13 \pm 0.10) + 0.10 \cdot \log I_{GS} \tag{4.10}$$

$$D = (470 \pm 50) + 26.3 \cdot \log I_{GS} \tag{4.11}$$

An example of the relationship of critical state angle ϕ'_c to I_{GS} is provided in Figure 4.8, distinguishing among the three soil classes and between drained (D) and undrained (U) tests. The critical angle of silts displays a larger range, reaching the upper values of sands, with scatter from the average value not exceeding $\pm 2°$, without any particular distinction among different soil classes and D/U tests.

It can be concluded that a well-defined distinction in the mechanical response between sands and silts was difficult to be assessed, whereas between SM-SP-ML and CL this distinction was clear. The one-dimensional and very small strain stiffness, both dependent on the current void ratio and effective stress level, could be clearly related to the grading properties through I_{GS} as well as the critical state parameters that showed a well-defined dependence on the grain size composition.

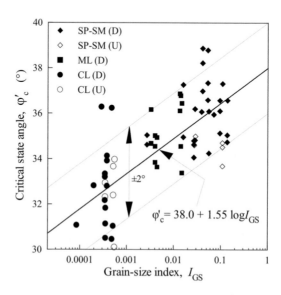

Figure 4.8 Critical state angle as a function of grain size index (Simonini, 2004).

From this point of view, I_{GS} may represent a helpful additional parameter to be used when evaluating the preliminary mechanical properties of highly heterogeneous and interbedded natural soils (e.g., Berengo et al. 2011). The sediments should, however, stem from a common mineralogical origin, such as in the case of the Venice lagoon.

4.3 THE TREPORTI TEST SITE

In order to measure directly on-site the soil stress–strain–time properties, a comprehensive research program was set up and completed in the first decade of the 21st century. The program involved the construction of a full-scale earth-reinforced cylindrical embankment over a typical soil profile in the Venice lagoon to load and measure relevant ground displacements and pore pressure evolution.

The trial embankment was built between September 2002 and March 2003 (a view of the resulting construction is provided in Figure 4.9). The height of the 40 m diameter cylindrical bank, realized using 13 superposed 0.5 m thick polypropylene geogrid-reinforced layers, was selected at 6.7 m to apply a uniform vertical stress of around 100 kPa (in reality equal to 106 kPa) to the ground surface. The sandy fill between each geogrid layer was dynamically compacted to give an average dry unit weight equal to 15.6 kN/m³.

The soil beneath the bank was continuously monitored to measure pore water pressure, surface settlements, and horizontal and vertical displacements with depth (for details, see Simonini 2004; Simonini et al. 2006). The monitoring continued for almost four years after construction as well as during the gradual removal of the embankment (from June 2007 to March 2008).

The bank area was extensively investigated by n. 10 CPTU (Tonni and Simonini 2013), n. 10 DMT (Mc Gillavry and Mayne 2004; Monaco et al.

Figure 4.9 Resulting construction of the Treporti trial embankment.

2014), n. 6 SCPTU and SDMT, and continuous coring boreholes and high-quality laboratory tests (Simonini 2004; Simonini et al. 2006). On the undisturbed samples standard classification and mechanical laboratory tests have been carried out. The DMT-SDMT and CPTU-SCPTU soundings were executed before starting the construction of the embankment, at the end of construction, from the top of the embankment, and after completing the gradual removal of the embankment.

4.3.1 Soil properties

Figure 4.10 depicts the basic soil properties determined on laboratory samples as a function of depth. The profile at TTS is similar to that at MTS with the exception of the absence of upper stiff OC silty clay, which had likely eroded during the last glaciation. The upper profile is composed of a medium-fine silty sand layer (2–8 m belowground level (g.l.)), located below a thin soft silty clay layer and followed by a relatively uniform thick silty layer (8–20 m), which was particularly important for the goal of the investigation. Below 20 m of depth, the soil is composed of alternating layers of clayey and sandy silt.

From soil grading measurement, the percentages of the various soil classes (up to 60 m) are as follows: SM-SP 22%, ML 32%, CL 37%, and CH-Pt 9%. The upper and deeper sands are relatively uniform, and finer materials are more graded. Excluding CH, Atterberg limits of the cohesive fraction are similar to those determined at MTS; in situ void ratio e_o is slightly higher with respect to that at MTS and lying approximately in the range between 0.8 and 1.1, with higher values due to laminations of organic materials.

The CPTU close to the embankment center provided the profile of Figure 4.10, last column on the right, in terms of relevant quantities q_t, f_s, and u_2 (cone resistance, sleeve friction, and pore pressure, respectively). Note the relevant flapping of u_2 confirming the large drainage differences in the whole deposit as a consequence of variations in grain size composition.

4.3.2 Monitoring

The instrumentation installed at TTS consisted of seven settlement plates at ground surface below the bank and 12 benchmarks outside the loaded area, 1 GPS antenna, 8 borehole rod extensometers, 4 special sliding deformeters (SD), 3 inclinometers, 5 Casagrande as well as 10 vibrating wire piezometers, and 5 load cells with total vertical stress beneath the loading embankment.

Figure 4.11a shows the planimetric position of the monitoring instrumentation, whereas Figure 4.11b depicts a cross section of the ground together with the trial embankment. A relevant role in interpreting the overall soil

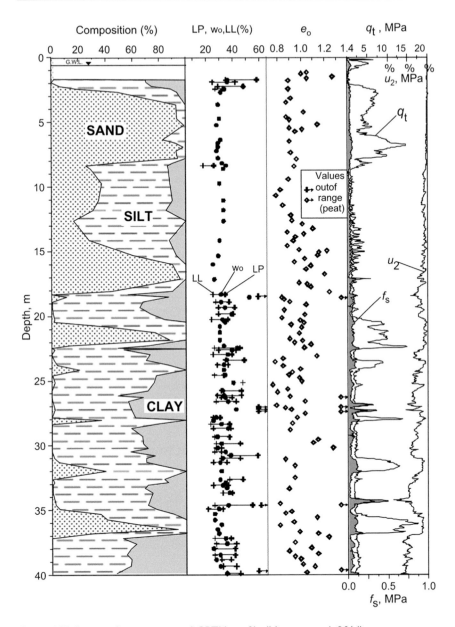

Figure 4.10 Basic soil properties and CPTU profile (Monaco et al. 2014).

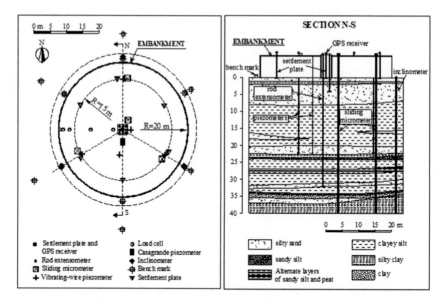

Figure 4.11 (a) Embankment plan with location of monitoring devices; (b) Cross section of embankment with soil profile and monitoring devices (Monaco et al. 2014).

behavior was provided by the four SD, to measure local vertical displacements of 1 m thick layers with a 0.03 mm/m degree of accuracy (Kovari and Amstad 1982).

Figure 4.12 depicts the evolution of the vertical displacement measured under the center of the embankment with the SD n. 3, the settlement plate n. 40, and the GPS. The total settlement on bank completion was 38.1 cm followed by an additional settlement at a constant load of 12.4 cm, thus giving a total of 50.5 cm. After bank removal, a very small settlement recovery was measured (<3.0 cm), confirming the extreme relevance of non-recoverable strain during the compression process of these silty-based soils.

It is worth noting that vertical displacements measured by three different methods during the entire six-year monitoring period, to a large extent, yielded coinciding results. A review of Figure 4.12 reveals that during the embankment construction a nearly entire consolidation settlement most likely occurred. This observation is confirmed by the piezometer readings (Simonini et al. 2006), showing that the excess pore pressure was negligible on bank completion and poorly distinguishable from the sea level oscillations induced by the daily tide in a nearby channel.

Figure 4.13 shows the settlement evolution of the ground surface at a half bank, on completion, after four years at constant load and after bank removal. The settlement is mostly concentrated below the bank with a small effect on the surrounding soil.

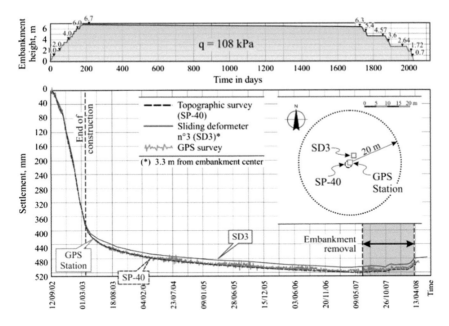

Figure 4.12 Time–settlement curves at the center of the trial embankment measured with different types of instruments (SD = sliding deformeter, SP = settlement plate) (Jiamiolkowski et al. 2009).

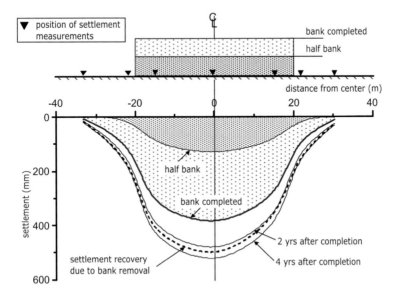

Figure 4.13 Settlements of the ground surface along a diameter section at different construction times (Monaco et al. 2014).

In addition, the maximum horizontal displacements measured with the inclinometers below the perimeter ($\approx$50 mm) were one order of magnitude lower than the maximum vertical settlements throughout the whole construction period, i.e., the deformation process developed prevalently in the vertical direction (Simonini 2004).

The SD n. 3 (and all the SD n. 1, 2, 4) provided measurements of local vertical strains of 1 m thick layers throughout the complete loading–unloading sequence. From Figure 4.14a and Figure 4.14b , which show, respectively, the local and total vertical strains with depth, the significant contribution provided by both the thin silty clay layer at $\approx$1–2 m depth and especially by the silt layer between 8 m and 20 m can be noted. The strain decreases with depth and is negligible below 35 m. Figure 4.15 depicts a few of the typical field compression curves. The vertical strain ε_v, measured in each 1 m thick layer by the SD n. 3, has been plotted vs. the vertical stress increment (starting from the geostatic vertical effective stress σ'_{v0}) estimated using the theory of elasticity solution for a uniformly loaded circular area. It is important to note the stiffer soil behavior at the first stress increments and the softer response beyond a threshold stress, which is more recognizable in silt rather than in sand. After bank completion, the deformation process was characterized by significant creep, followed by a very stiff unloading response.

Since strain in the ground developed prevalently vertically, it was assumed that the curves in Figure 4.15 may be viewed as a sort of 'one-meter field oedometer' curves and that the threshold stress may be considered as the preconsolidation stress σ'_{vy}. However, as discussed above, the 'one-meter field oedometers' are not, technically, oedometers, because the various stress paths upon loading are below/above the K_0 line. These sets of data made possible the estimation of σ'_{vy} with depth, thus assessing the OCR = $\sigma'_{vy}/\sigma'_{v0}$ profile (shown in Figure 4.16). This estimate of OCR is particularly crucial, since it is the first attempt to precisely measure directly on-site the preconsolidation/precompression stress of Venetian soils, in both silts and sands.

Note that the soils are characterized by OCR between 1.5 and 2.5 for the shallower Holocenic soils (over the *caranto* layer) and less than 1.3 for the Pleistocenic sediments (under the *caranto* layer, as already described in Chapter 2).

Figure 4.17 reports typical vertical displacement vs. log(time) curves evaluated for each 1 m thick layer. The strain–time trend is characterized by an *S-type* shape with the final part, corresponding to deformation occurring after bank completion, fitted plane by a straight line (compression coefficient $C_{\alpha\,\varepsilon\text{-site}} = \Delta\varepsilon_v/\Delta\log t$). The curves show very similar shapes to those obtained by plotting vertical displacement readings vs. log$t$ in an oedometer test for a given load increment, thus allowing the classical and simplified interpretation in terms of primary and

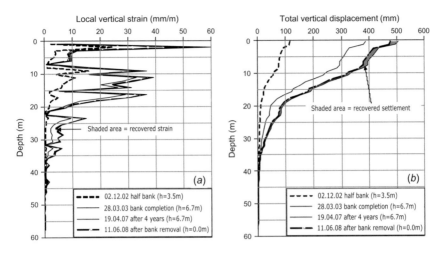

Figure 4.14 (a) Local vertical strain and (b) total vertical displacement measured by sliding deformeter (SD) n. 3 (Jamiolkowski et al. 2009).

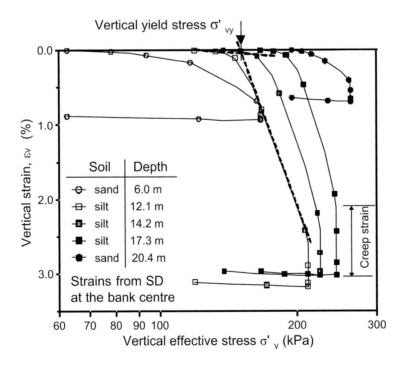

Figure 4.15 Typical field compression curves ('one-meter field oedometer') in sands and silts (Monaco et al. 2014).

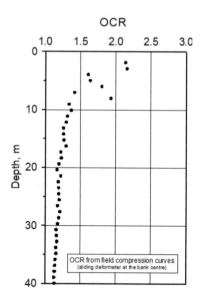

Figure 4.16 Overconsolidation ratio from field stress–strain curves (Monaco et al. 2014).

secondary deformation stages. The approximately linear trend observed in the last part of each curve is defined by the deformation data measured after the bank construction; hence, it is very likely that vertical strain due to primary consolidation developed simultaneously with the loading process.

4.3.3 Stiffness and compressibility from site monitoring

From the strain measurements provided by the sliding deformeters, it was possible to estimate some mechanical soil properties of all the types of soil encountered at TTS, thus avoiding both the scale effect and the stress relief and/or destructuration due to sampling, particularly affecting the mechanical response of silts. It was also possible to evaluate compression stiffness for sandy materials, as well. To this purpose, a linear elasticity-based vertical stress distribution induced by the trial embankment was assumed, as discussed above.

The following relevant parameters have been determined:

- Initial stiffness $M_{i\text{-site}}$, evaluated as the slope of the tangent to the vertical stress σ'_v vs. vertical strain ε_v curves for the central SD n. 3;
- Current stiffness $M_{site} = \Delta \sigma'_v/\Delta \varepsilon_v$ evaluated for each loading step for the central SD n. 3 in the silty layer at depths from 8 m to 20 m;

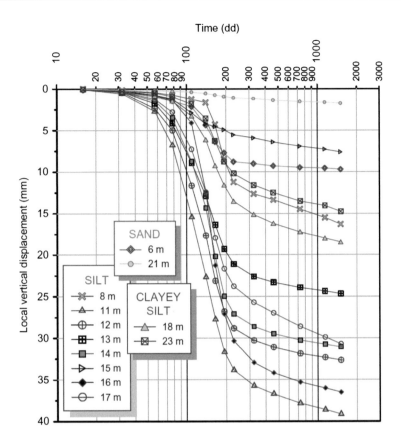

Figure 4.17 Curves of the local vertical displacements measured beneath the center of the loaded area vs. time, at different depths. In both plots, time is expressed in terms of days since the bank initial loading (Tonni and Simonini 2013).

- Primary compression index C_{ce} defined as the slope of $\varepsilon_v - \log\sigma'_v$ curves beyond estimated yielding stress σ'_{vy} for the SD n. 3 and for the SD n. 1, 2, and 4;
- Secondary compression index $C_{\alpha\varepsilon}$ defined as the slope of the $\varepsilon_v - \log t$ curves after the end of embankment construction for SD n. 3 and for SD n. 1, 2, and 4.

The initial stiffness $M_{\text{i-site}}$ is plotted against depth in Figure 4.18 together with the maximum shear stiffness measured with SCPTU and transformed into elastic modulus in both confined M_{SCPTU} and unconfined conditions E_{SCPTU} (assuming a Poisson's ratio = 0.20). As expected, $M_{\text{i-site}}$ is characterized by values lower than the SCPTU stiffness profile, except for a few shallower layers, where $M_{\text{i-site}}$ seems to approach the seismic stiffness.

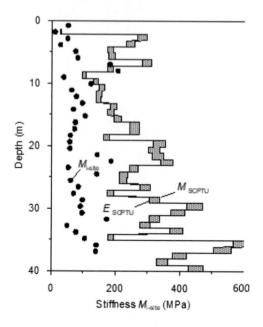

Figure 4.18 Comparison between initial stiffness and maximum stiffness from seismic piezocone (Simonini et al. 2006).

Current stiffness was calculated for all the layers and, after normalizing against the vertical effective stress $(M/\sigma\,'_v)_{site}$, was plotted vs. the current vertical stress $(\sigma\,'_v/\sigma\,'_{vy})_{site}$ induced by the increasing embankment load. Figure 4.19 reports the trend of stiffness against current vertical stress only for the sandy silt layer at depths between 8 and 20 m, which is the most responsible for the occurred settlements (called Layer B).

It is interesting to notice the sharp variation of $(M/\sigma\,'_v)_{site}$ trend before and after the preconsolidation stress that clearly separates the behavior in OC from normally consolidated (NC) range. The $(M/\sigma\,'_v)_{site}$ values are compared in the same Figure 4.19 with the laboratory data of the normalized constrained modulus $(M/\sigma\,'_v)_{lab}$ obtained from oedometric compression tests. In order to compare data as homogeneous as possible, $(M/\sigma\,'_v)_{lab}$ was estimated from the interpretation of the recompression curve in an unloading–reloading cycle followed by the virgin compression part (dashed line shown in the upper-right part of Figure 4.19). With this procedure, the normalizing yielding stress is the maximum vertical stress before an unloading–reloading cycle, that is, the precompression effect on the response in oedometric tests is due to only mechanical imposed precompression, thus reducing the possible intrinsic influence of aging or structuration on soil stiffness.

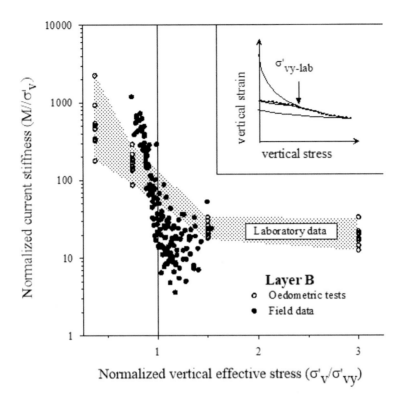

Figure 4.19 Site and laboratory normalized tangent stiffness vs. normalized vertical stress (Simonini et al. 2006).

It is important to point out that field data cross the laboratory trend; the stiffness before yielding stress is significantly higher compared to laboratory data, whereas beyond $(\sigma\,'_v/\sigma\,'_{vy})_{lab}$ it reduces to significantly lower values. This may be explained by considering the effect of disturbance and stress relief due to sampling. From the above considerations, it seems that the small precompression observed in Pleistocene formations at TTS should be due to the combined effect of mechanical precompression and aging, not excluding the possible influence of an extremely slight structuration.

The trend of site compression curves was interpreted in terms of virgin compression coefficient $C_{c\varepsilon}$ as well as of secondary compression coefficient $C_{\alpha\varepsilon}$. Both are plotted in Figure 4.20a and b as a function of depth.

Despite a certain scatter of results, mostly due to high heterogeneity and chaotic alternance of different sediments within the 1 m interval, it was found that typical values of $C_{c\varepsilon}$ for sands generally fall in the interval 0.04–0.05, whereas higher values of $C_{c\varepsilon}$ were provided by the silty layer from 8 m to 20 m depth, ranging from 0.09 to 0.34. The highest values (0.23–0.34) were computed below 15 m depth, where the clay content in

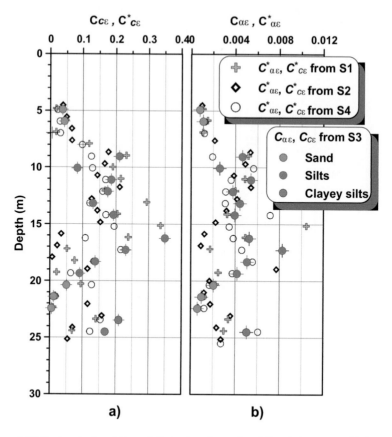

Figure 4.20 Back-calculated $C_{c\varepsilon}$ and $C_{c\varepsilon}$* values from the sliding deformeters (SD) n. 1, 2, 3, and 4; (b) back-calculated $C_{\alpha\varepsilon}$ and $C_{\alpha\varepsilon}$* values from all the sliding deformeters (Tonni and Simonini 2013).

the silty sediments becomes significant. The procedure was also applied to the vertical strain ε_v vs. $\log\sigma'_v$ curves associated with the other SD n. 1, 2, and 4 (located at 120° along with a 15 m radius circumference, see Figure 4.11a), and a compression ratio $C_{c\varepsilon}$*, though not strictly referable to one-dimensional conditions, was obtained.

Secondary compression coefficient $C_{\alpha\varepsilon}$ was derived from data associated with the early stages of the secondary compression process. In this way, $C_{\alpha\varepsilon}$ in silts varies between 2.6×10^{-3} and 5.4×10^{-3}, with a mean value of approximately 4.6×10^{-3}. Secondary compression in sands is typically described by $C_{\alpha\varepsilon}$ in the range 5.8×10^{-4} to 9.5×10^{-4}. A modified coefficient of secondary compression $C_{\alpha\varepsilon}$* has also been calculated from SD n. 1, 2, and 4, and the results are plotted in Figure 4.20b.

4.4 SOIL PROPERTIES FROM CPTU AND DMT

The excellent database provided by the Malamocco and Treporti test sites allowed evaluation of the applicability of CPTU/SCTPU and DMT/SDMT for extensive soil characterization at the three inlets and in the whole lagoon, including some areas on the closest mainland. Details of such evaluation can be found in Simonini (2004), Simonini et al. (2006), Tonni and Simonini (2013), Monaco et al. (2014), and Tonni et al. (2016). Here only a couple of examples of estimating the mechanical parameters are reported (one from CPTU and the other from DMT) relevant to settlement prediction, which is the most difficult goal to be achieved for these highly heterogeneous silty soils.

4.4.1 Estimate of secondary compression of silts and sands from CPTU

Field observations discussed in the previous section have shown that time-dependent behavior is not negligible in both Venetian silts and sands; hence, the proper evaluation of the secondary compression parameter is of crucial importance for long-term settlement predictions of all the structures lying on the Venetian soils.

Due to the predominantly silty and highly heterogenous nature of such sediments, undisturbed sampling is rather difficult; hence, any possibility of deducing reliable compression parameters around the geostatic effective stress state from laboratory tests on specimens affected by scale effect and stress relief due to sampling is excluded.

The attractive idea of using CPTU to estimate secondary compression in Venetian silts and sands is mainly based on the observation that in these soils cone resistance and secondary compression are basically dependent on friction, therefore providing valid reasons to explore the occurrence of a relationship between the two quantities (Tonni and Simonini 2013, Tonni et al. 2016). In other words, since in these soils both cone penetration and creep behavior are basically governed by the frictional response, there are valid reasons to expect an effective relationship between $C_{\alpha \varepsilon}$ and the cone resistance.

Tonni and Simonini (2013) developed a power function expression between $C_{\alpha \varepsilon}$ and dimensionless normalized cone resistance Q_{tn}, as defined by Robertson (2009). In this approach, an iterative, nonlinear stress normalization procedure is applied to the corrected cone resistance q_t, leading to the following expression:

$$Q_{tn} = \frac{q_t - \sigma_{z0}}{p_a} \cdot \left(\frac{p_a}{\sigma'_{v0}} \right)^n \tag{4.12}$$

where p_a is the atmospheric pressure and n is a stress exponent that depends in turn on both stress level and the well-known soil behavior type index I_c, originally defined as:

$$I_c = \sqrt{(3.47 - \log Q_t)^2 + (\log F_r + 1.22)^2} \tag{4.13}$$

with

$$F_r = 100 \cdot \frac{f_s}{q_t - \sigma_{v0}} \tag{4.14}$$

$$Q_t = \frac{q_t - \sigma_{v0}}{\sigma'_{v0}} \tag{4.15}$$

According to experimental observations, Robertson (2009) suggested the following expression for n:

$$n = 0.38 I_c + 0.05 \cdot (\sigma'_{v0} / p_a) - 0.15 \tag{4.16}$$

where the soil behavior type index I_c should be defined using Q_{tn} instead of Q_t, hence the iterative nature of the scheme.

In this way, a variable exponent n (≤ 1) is adopted for stress normalization, depending on the soil class to be considered. This exponent turned out to vary between 0.5 and 0.6 in the upper sandy layer, while in the silty unit it tended toward 1.0.

The effectiveness of the normalization procedure has been confirmed with respect to the TTS piezocone database.

Figure 4.21 shows the profile of the computed I_{cn} (expressed in terms of Q_{tn}) deduced from CPTU 14, carried out at the center of the bank, in conjunction with the I_{cn}-based boundaries of the soil type zones, identified by Robertson on the well-known normalized *soil behavior type* (SBT) Q_t–F_r chart and later included in the recent SBTn Q_{tn}–F_r chart (Robertson 2009). Furthermore, the profile of the normalized cone resistance Q_{tn} was plotted.

The relationship between $C_{\alpha\varepsilon}$ (including $C_{\alpha\varepsilon}{}^*$ values) and Q_{tn} is presented in Figure 4.22 where about 70 paired observations of $C_{\alpha\varepsilon}$ and Q_{tn}, obtained from all sliding deformeters and piezocone data are displayed. Points associated with secondary compression and cone resistance recorded along the centerline have been depicted using different symbols, in relation to the different grain size characteristics of sediments.

Despite a certain scatter, plotting data in a log-log plane provides evidence of a rather straightforward trend between $C_{\alpha\varepsilon}$ and Q_{tn}, which can be described by a power function of the form:

$$C_{\alpha\varepsilon} = 0.03 \cdot (Q_{tn})^{-0.89} \tag{4.17}$$

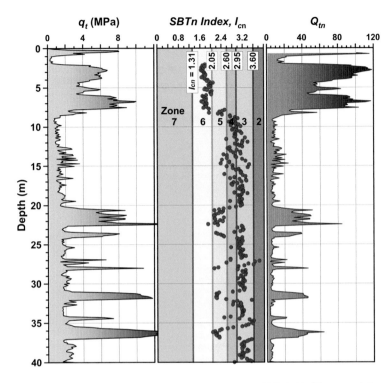

Figure 4.21 Profile of the computed I_{cn} deduced from CPTU 14 (Tonni and Simonini 2013).

with $R^2 = 0.83$.

Considering all the available data, the general regression trend is given by:

$$C_{\alpha\varepsilon} = 0.018 \cdot \left(Q_{tn}\right)^{-0.69} \tag{4.18}$$

with $R^2 = 0.68$. In this latter case, $C_{\alpha\varepsilon}$ stands for both $C_{\alpha\varepsilon}$ and $C_{\alpha\varepsilon}{}^*$, according to previous definitions.

Despite the satisfactory predictive capability of the above correlations, it is worth mentioning that neither of them explicitly considers partial drainage phenomena during cone penetration in Venetian silts. Such an effect should be carefully assessed in order to improve understanding and interpretation of penetration data in intermediate soils.

It therefore seemed reasonable to include an additional factor in the Q_{tn}–$C_{\alpha\varepsilon}$ relationship, accounting for the different pore pressure responses in relation to the drainage conditions around the advancing cone. Following recent studies on trends in normalized piezocone response as a function of the drainage degree (e.g., Schneider et al. 2007, 2008), the excess pore

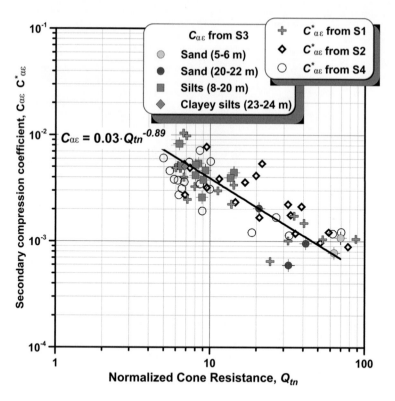

Figure 4.22 Relationship between $C_{\alpha\varepsilon}$ and Q_{tn} (Tonni and Simonini 2013).

pressure ratio, $\Delta u / \sigma'_{v0}$, was eventually considered as the simplest and most appropriate variable to adopt in order to solve the issue.

Thus, a multiple regression analysis was performed in a log-log format to provide a power function expression for $C_{\alpha\varepsilon}$ in terms of the normalized cone tip resistance Q_{tn} and the excess pore pressure ratio-based factor $(1 + \Delta u / \sigma'_{v0})$.

If the analysis is restricted to include only the data collected over the centerline of the test bank, the equation of the best-fit regression curve turns out:

$$C_{\alpha\varepsilon} = 0.077 \cdot \left(Q_{tn}\right)^{-1.14} \cdot \left(1 + \frac{\Delta u}{\sigma'_{z0}}\right)^{-0.74} \qquad (4.19)$$

with $R^2 = 0.86$.

For all the available data, the least square analysis gave the following relationship:

$$C_{\alpha\varepsilon} = 0.035 \cdot \left(Q_{tn}\right)^{-0.87} \cdot \left(1 + \frac{\Delta u}{\sigma'_{v0}}\right)^{-0.55} \qquad (4.20)$$

with $R^2 = 0.72$.

It is interesting to note that the use of $(1 + \Delta u/\sigma'_{v0})$ gives the highest R^2, whether the analysis includes only the centreline-located data or all the available measurements, thus confirming that this specific factor and $C_{\alpha\varepsilon}$ can be related in a rational manner.

The above equations were satisfactorily used to back-analyze and predict the long-term settlement of existing structures for the protection of the lagoon. Details of these applications can be found in Tonni et al. (2016).

4.4.2 Estimate of OCR for sands and silts from DMT and CPTU

The previous section emphasized the difficulty of estimating preconsolidation/precompression stress from oedometric tests due to the lack of a clear threshold value that separates the overconsolidated from normally consolidated states of silty-based materials. OCR provides fundamental information for a suitable settlement prediction.

However, the test embankment program provided an estimate of the OCR profile, according to the definition of OCR, at two times, that is, at full load and after load removal, shown in Figure 4.23.

Site investigations were carried out not only before the beginning of the embankment construction but also at the end of construction (Site Investigation 2: SI-2), from the top of the embankment (2003), and after

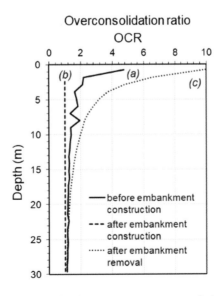

Figure 4.23 OCR profile before bank construction, at completion, and after removal (Monaco et al. 2014).

completing the gradual removal of the embankment (Site Investigation 3: SI-3) (2008).

At full load it was considered OCR $\approx$ 1, assuming that at every depth the vertical stress had exceeded the previous maximum past pressure. After load removal, the OCR was evaluated assuming σ'_{vmax} as the geostatic stress plus the vertical stress increment induced by the uniformly loaded circular area, according to the theory of elasticity. Figure 4.21 also includes (curve *a*) a tentative OCR profile of the undisturbed soil before loading, based on the maximum curvature points in Figure 4.15.

Correlations between OCR and DMT in sand have been attempted by Schmertmann (1983), Marchetti (1985), and Mayne et al. (2009). Currently, the method that is considered generally the most applicable, although highly approximate, is the method described in TC16 (2001). This method makes use of the ratio between the constrained modulus M from DMT (M_{DMT}) and the cone penetration resistance q_t from CPTu. Semi-quantitative guidelines reported in TC16 (2001) are as follows: M_{DMT}/q_t = 5–10 in NC sands, M_{DMT}/q_t = 12–24 in OC sands.

As is widely known, the α factor by which the tip resistance q_t must be factorized, in order to obtain an estimate of the 'operative' Young's modulus E', increases significantly with OCR. Extensive calibration chamber research on sands (e.g., Bellotti et al. 1989) has indicated typical α values $\approx$3–4 in NC sand and up to $\approx$20 in OC sands. Since the ratio α increases with OCR, it may be used as an indicator of OCR. Since moduli increase with OCR at a faster rate than the penetration resistance, the ratio between modulus and penetration resistance should increase with OCR.

Figure 4.24 shows the correlation OCR-M_{DMT}/q_t for the TTS sands (modified from Monaco et al. 2014). It was constructed using same-depth values of M_{DMT} and q_t obtained from three soundings in sandy layers (having material index I_D > 1.8) between 2 m and 35 m depth.

The DMT/CPTU data were those obtained at the times when the reference OCR profiles were available (curves *b* and *c* in Figure 4.23), i.e., at bank completion and post-removal. The data pairs M_{DMT}-q_t were carefully selected to avoid any possible mismatching of data, by retaining only pairs from uniform soil layers of significant thickness. The equation of the interpolating line is:

$$OCR = 0.3051\,(M_{DMT}/q_t) - 0.9975 \quad (4.21)$$

The OCR-M_{DMT}/q_t data points in Figure 4.24 are in good agreement with the TC16 (2001) guidelines (M_{DMT}/q_c = 5–10 in NC sands and M_{DMT}/q_c = 12–24 in OC sands). Considering that Equation (4.21) is in satisfactory agreement with the preexisting experimental base, it may tentatively be assumed that the equation can provide a broad OCR estimate for sands and silts in the Venetian lagoon.

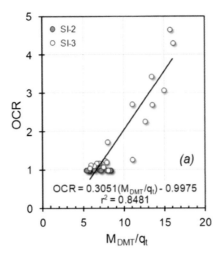

Figure 4.24 Correlation between OCR and M_{DMT}/q_t (Monaco et al. 2014).

All the geotechnical investigations carried out at MTS and TTS provided information from which it was possible to interpret the geotechnical data and provide a framework that describes the stress–strain–time behavior of non-textbook heterogeneous silty soils, considering their significant heterogeneity as well and, therefore, the inherent scale effect.

REFERENCES

Bellotti, R., Ghionna, V., Jamiolkowski, M. and Robertson, P. K. 1989. Shear strength of sand from CPT. *Proceedings 12th International Conference on Soil Mechanics and Foundation Engineering* 1, 179–184.

Berengo, V., Benz, T., Simonini, P. and Leoni, M. 2011. Site monitoring and numerical modelling of a trial embankment's behaviour on Venice Lagoon soils. *International Scholarly Research Notices* 2011, 11, Article ID 378579. https://doi.org/10.5402/2011/378579.

Biscontin, G., Pestana, J., Cola, S. and Simonini, P. 2001. Influence of grain size on the compressibility of Venice Lagoon soils. *XV ICSMGE* 1, 35–38.

Biscontin, G., Pestana, J., Cola, S. and Simonini, P. 2007. A unified compression model for the Venice lagoon natural silts. *Journal of Geotechnical and Geoenvironmental Engineering* 133(8), 932–942.

Bolton, M. D. 1986. The strength and dilatancy of sands. *Géotechnique* 36(1), 65–78.

Bolton, M. D. 1987. Discussion. *Géotechnique* 37(2), 225–226.

Cola, S. and Simonini, P. 2002. Mechanical behaviour of silty soils of the Venice lagoon as a function of their grading properties. *Canadian Geotechnical Journal* 39(4), 879–893.

Hardin, B. O. and Black, W. L. 1969. Vibration modulus of normally consolidated clay. *Journal of Soil Mechanics & Foundations Division* 95(1), 33–65.

Hardin, B. O. and Drnevich, V. P. 1972. Shear modulus and damping in soils: Design equations and curves. *Journal of the Soil Mechanics and Foundations Division* 98(7), 667–692.

Ishihara, K., Tatsuoka, F., and Yashuda, S. 1975. Undrained strength and deformation of sand under cyclic stresses. *Soils and Foundations* 15(1), 29–44.

Jamiolkowski, M., Ricceri, G. and Simonini, P. 2009. Safeguarding Venice from high tides: Site characterization and geotechnical problems. Keynote lecture. In *Proceedings of the 17th International Conference on Soil Mechanics and Geotechnical Engineering* (Volumes 1, 2, 3 and 4). October 4–9. Alexandria: IOS Press, 3209–3227. From the pre-print version available at UNIPD IRIS website https://www.research.unipd.it/item/preview.htm?uuid=4b48dac1 -44ac-4dd9-9e4f-9b11b7dc7395.

Janbu, N. 1963. Soil compressibility as determined by oedometer and triaxial tests. *Proceedings European Conference S. M. F. E. Wiesbaden* 1, 19–25.

Kovari, K. and Amstad, C. 1982. A new method of measuring deformations in diaphragm walls and piles. *Géotechnique* 22(4), 402–406.

Mesri, G. and Godlewski, P. M. 1977. Time- and stress- compressibility relationship. *Journal of Geotechnical and Geoenvironmental Engineering ASCE* 105(5), 417–430.

Mesri, G., Shahien, M. and Feng, T. W. 1995. Compressibility parameters during primary consolidation. *Proceedings International Symposium on Compression and Consolidation of Clayey Soils* 2, 1021–1037.

Marchetti, S. 1985. On the field determination of k_0 in sand. *International Conference on Soil Mechanics and Foundation Engineering* 11, 2667–2672.

Mayne, P. W., Coop, M. R., Springmanm, S. M., Huang, A. and Zornberg, J. G. 2009. Geomaterial behavior and testing. In M. Hamza et al. (Eds.), *Proceedings of 17th International Conference on Soil Mechanics and Geotechnical Engineering*. Alexandria: IOS Press, 2777–2872.

McGillivray, A. and Mayne, P. W. 2004. Seismic piezocone and seismic flat dilatometer tests at Treporti. In A. Viana da Fonseca and P. W. Mayne (Eds.), *Proceedings of 2nd International Conference on Site Characterization ISC'2, Porto* (Vol. 2). Rotterdam: Millpress, 1695–1700.

Monaco, P., Amoroso, S., Marchetti, D., Marchetti, S., Totani, G., Cola, S. and Simonini, P. 2014. Overconsolidation and stiffness of Venice Lagoon sands and silts from SDMT and CPTU. *Journal of Geotechnical and Geoenvironmental Engineering ASCE* 140(1), 215–227.

Roberston, P. K. 2009. Interpretation of cone penetration tests – A unified approach. *Canadian Geotechnical Journal* 46, 1337–1355.

Schmertmann, J. H. 1983. Revised procedure for calculating k0 and OCR from DMT's with ID > 1.2 and which incorporates the penetration force measurement to permit calculating the plane strain friction angle. DMT Digest No.1, GPE Inc., Gainesville, FL.

Schneider, J. A., Lehane, B. M. and Schnaid, F. 2007. Velocity effects on piezocone measurements in normally and over consolidated clays. *International Journal of Physical Modelling in Geotechnics* 2, 23–34.

Schneider, J. A., Randolph, M. F., Mayne, P. W. and Ramsey, N. R. 2008. Analysis of factors influencing soil classification using normalized piezocone tip resistance and pore pressure parameters. *Journal of Geotechnical and Geoenvironmental Engineering ASCE* 134(11), 1569–1586.

Simonini, P. 2004. Characterization of the Venice lagoon silts from in-situ tests and the performance of a test embankment. Keynote Lecture. In A. Viana da Fonseca and P. W. Mayne (Eds.), *Proceedings 2nd International Conference on Site Characterization ISC'2, Porto* (Vol. 1). Rotterdam: Millpress, 187–207.

Simonini, P., Ricceri, G. and Cola, S. 2006. Geotechnical characterization and properties of Venice lagoon heterogeneous silts. Invited lecture. Characterization and Engineering Properties of Natural Soils. 29 November – 1 December 2006, Singapore, published as preprint in http://hdl.handle.net/11577/2443280.

Tatsuoka, F. and Ishihara, K. 1974. Drained deformation of sand under cyclic stress reversing direction. *Soils and Foundations* 14(3), 51–65.

ISSMGE committee TC16. 2001. The Flat Dilatometer Test (DMT) in soil investigations – A report by the ISSMGE committee TC16. In R. A. Failmezger and J. B. Anderson (Eds.), *Proceedings 2nd International Conference on the Flat Dilatometer*. Washington, DC, 7–48.

Tonni, L. and Simonini, P. 2013. Evaluation of secondary compression of sands and silts from CPTU. *Geomechanics and Geoengineering* 8(3), 141–154.

Tonni, L., García Martínez, M. F., Simonini, P. and Gottardi, G. 2016. Piezocone-based prediction of secondary compression settlements of coastal defence structures on natural silt mixtures. *Ocean Engineering* 116, 101–116.

Chapter 5

Venetian soil constitutive modeling

Written with Simonetta Cola

5.1 INTRODUCTION

Chapter 4 has presented the main features of the Venetian soils that are prevalently composed of silty sediments mixed with sand and/or clay according to highly heterogeneous soil profiles, rapidly variable even in closely spaced sites, and associated scale effects.

Constitutive modeling of these intermediate soils, useful to predict their short- and long-term behavior when loaded by external loads, is therefore challenging.

Some classical models have been applied to simulate the response of Venetian soils under loading, basically those implemented in commercial codes (e.g., Cortellazzo et al. 2007), or more advanced material models such as those used to predict the settlements under the caissons foundations of the mobile barriers under construction at the inlets, as described in detail in Chapter 6. In this case, the settlement of coarse-grained materials was computed assuming a fully drained loading and adopting the elasto-plastic strain hardening model proposed by Li and Dafalias (2000); the time-settlement behavior of fine-grained soils was evaluated by a viscous-plastic coupled consolidation soil model adapted to the conditions of one-dimensional compression by Rocchi et al. (2003).

The constitutive models to be applied require sketched ground profiles, within which the layering is clearly defined distinguishing between coarse and fine-grained soils. In fact, soil models have been developed to properly describe textbook soils such as sands or clays (e.g., those cited above) but not intermediate soils such as silts or mixtures of silts with clay or sand.

This distinction between clay and sand, which is necessary when using numerical approaches, may be relatively difficult in the case of Venetian silts, due to their highly heterogeneous nature.

Biscontin et al. (2001, 2007) presented a new one-dimensional constitutive approach specifically developed for Venetian heterogeneous silt. This approach has proven to properly replicate the behavior of the subsoil under the embankment at the *Treporti test site*, time effects included. This model is an advancement of the one proposed by Pestana and Whittle (1995,

DOI: 10.1201/9781003195313-5

1998), who developed a compression model for both sands and sandy silts that was able to describe the nonlinear compression response over a wide range of confining stresses and densities. Pestana and Whittle (1999) suggested that the same model may be used to describe the one-dimensional compression behavior of soils containing a mix of compressible (clay) and incompressible (sand and silt) materials. More details on the model can be found in Biscontin et al. (2007).

5.2 CONSTITUTIVE EQUATIONS

5.2.1 Elastoplastic behavior

The hypotheses on which the one-dimensional elastoplastic model is based are the following:

- During the one-dimensional compression process, the change of the volumetric strain $d\varepsilon_p$ can be decomposed into elastic $d\varepsilon_p^e$ and plastic $d\varepsilon_p^p$ components;
- The elastic modulus (either volumetric or shear modulus) to calculate elastic strain is a function of the current mean effective stress p' and of the void ratio e;
- The plastic modulus to calculate plastic component depends on the distance between the current vertical effective stress and a unique limit compression curve in one-dimensional condition (K_0–LCC), approached at higher stress levels by all the soil samples despite starting from different initial formation void ratios;
- A unique limit compression curve (K_0–LCC) at higher stress levels exists with its slope ρ_c uniquely dependent on the specific soil mineralogy and constant over time.

Figure 5.1 shows K_0–LCC in a double logarithmic void ratio-effective stress space. It is completely described by the slope, ρ_c, and the location given by the reference vertical stress at a void ratio of one, σ'_{vr}, or the reference void ratio at a stress of 1 atm, e_{1v}, given by:

$$e = \left(\frac{\sigma'_v}{\sigma'_{vr}}\right)^{-\rho_c} = e_{1v}\left(\frac{\sigma'_v}{p_{at}}\right)^{-\rho_c} \tag{5.1}$$

where e is the current void ratio, σ'_v is the vertical effective stress, and p_{at} is the atmospheric pressure.

In the one-dimensional compression process, the change of the volumetric strain $d\varepsilon_p$ can be decomposed into elastic $d\varepsilon_p^e$ and plastic $d\varepsilon_p^p$ components. The elastic component $d\varepsilon_p^e$ may be determined from:

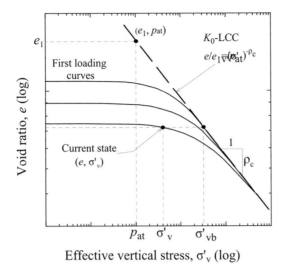

Figure 5.1 Idealized one-dimensional compression response of sands with different initial densities (Biscontin et al. 2007).

$$d\varepsilon_p^e = \frac{dp'}{K^e} = \frac{3(1-2v)}{(1+v)}\frac{1}{G^e}\left(\frac{1+2K_0}{3}\right)d\sigma'_v \qquad (5.2)$$

where (σ'_v, e) is the current vertical effective stress and void ratio, respectively; K_0 is the coefficient of lateral earth pressure at rest; K^e and G^e are the elastic bulk and shear modulus; and v is the elastic Poisson's ratio.

The elastic shear stiffness, also referred to as G^e (the maximum elastic stiffness), can be described by the following relationship:

$$G^e = \frac{p_{atm}G_b}{e^m}\left(\frac{p'}{p_{atm}}\right)^n \qquad (5.3)$$

where G_b is a material constant, and m and n are two exponents.

Plastic $d\varepsilon_p^p$ deformation is controlled by the proximity of the current stress σ'_v to the corresponding vertical stress σ'_{vb} on the K_0–LCC at the same void ratio.

$$d\varepsilon_p^p = \frac{e}{1+e}\rho_c\left(1-\delta_b^\vartheta\right)\frac{d\sigma'_v}{\sigma'_v} \qquad (5.4)$$

$$\delta_b = 1-\sigma'_v/\sigma'_{vb} \qquad 0\le\delta_b\le 1 \qquad (5.5)$$

$$\frac{\sigma'_{vb}}{p_{at}} = \left(\frac{e_{1v}}{e}\right)^{1/\rho_c} \tag{5.6}$$

where θ is a model parameter controlling the curvature of the compression curve and δ_b is the normalized distance between the current vertical effective stress and the vertical effective lying on the K_0–LCC at the current void ratio.

The same model could be used to describe the one-dimensional compression response of transitional soils, namely saturated soils containing a mix of cohesive and granular soils such as those of the Venice lagoon.

According to Mitchell (1976), these soils can be idealized as a combination of two phases: (Figure 5.2), the clay–water phase, also referred to as the clay matrix, and the granular phase, namely the silt and sand fraction.

By assuming for practical purposes that the specific weight of clay, silt, and sand particles is the same, the volumetric clay fraction f_c ($0 \leq f_c \leq 1$) or the gravimetric clay fraction FF ($0 \leq FF \leq 1$) is given by:

$$FF = \frac{W_c}{W_c + W_g} \cong \frac{V_c}{V_c + V_g} = f_c \tag{5.7}$$

The fine fraction FF, separating finely and coarsely grained materials, is customarily the percent by weight smaller than 2 µm; however, in the case of Venetian soils, this value is assumed to be slightly higher, that is 5 µm. For the Venetian silty clays, the value of 5 µm does in fact appear to take the separation into account, better than the classical value of 2 µm.

The void ratios of the clay–water and granular phases, e_c and e_g, respectively, can be determined from the overall void ratio e and the volumetric clay fraction, according to the following relationships:

$$e_c = \frac{V_w}{V_c} \approx \frac{e}{FF} \tag{5.8}$$

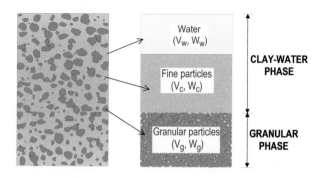

Figure 5.2 Bi-phase conceptual model (Biscontin et al. 2007).

$$e_g = \frac{V_w + V_c}{V_g} \approx \frac{e + FF}{1 - FF} \qquad (5.9)$$

According to Equation (5.9), the value of e_g is a function of water content and clay fraction FF. When e_g is higher than the maximum void ratio, $e_{g,max}$, of the granular phase alone, the grains behave as rigid inclusions in the clay–water phase and provide insignificant contribution to the overall compressibility, the latter being controlled only by the clay matrix.

The location of the K_0–LCC becomes a function of the clay fraction:

$$e_{1v} = e_{c1v} \cdot FF \qquad (5.10)$$

where e_{c1v} is the reference void ratio for the K_0–LCC of the clay–water phase only.

The reference void ratio e_{c1v} for the clay–water phase and the slope, ρ_c, of the K_0–LCC line are uniquely dependent on the specific mineralogy of the clay phase.

For soil samples with different FF but same clay mineralogy, the location of the K_0–LCC lines changes according to Equation (5.10), but the slope, ρ_c, remains unchanged. If these compression curves are normalized by their respective e_{1v}, they converge to the same normalized K_0–LCC, independent from FF.

On the contrary, if the saturated clay fraction is small, the granular particles are in mutual contact, and the clay matrix is entrapped in the voids of the granular skeleton. In this case, the overall compressibility is controlled by the granular phase alone.

The overall void ratio is therefore related to FF and e_g by Equation (5.9), and the one-dimensional compression curve becomes a straight line, characterized by the slope ρ_c of the granular fraction.

The K_0–LCC line position changes according to:

$$e_{1v} = e_{g1v}(1 - FF) - FF \qquad (5.11)$$

where e_{g1v} is the reference void ratio for the K_0–LCC of the granular phase only. Normalizing each compression curve by its reference void ratio e_{1v} results, again, in convergence to the normalized K_0–LCC, independent from clay fraction.

For intermediate soil, the transition from a mechanical response controlled by the coarse fraction toward one totally governed by the fine fraction occurs gradually, in a relatively large FF range and not for an FF threshold value, as predicted by the intersection of Equations (5.10) and (5.11). In this case e_{1v} is a function of e_{c1v}, e_{g1v}, and FF and will be defined in the next section.

On the basis of the above equations, the model is completely defined by seven independent parameters, namely three parameters for the elastic strain component (G_b, n, and ν) and three for the plastic component (e_{1v}, ρ_c, and θ).

5.2.2 Incorporation of viscous effect

With small modifications, the above elastoplastic model may be adapted to incorporate viscous effects, which are quite substantial in Venetian soils. In this case, the plastic component of strain is considered viscoplastic $d\varepsilon_p^{vp}$; in other words, it also takes the time-dependent component of the irrecoverable strain into account.

To this purpose, the K_0–LCC depicted with continuous line and determined for a given soil using an oedometric test carried out applying constant load increments may be assumed to be related to a reference time t_{ref}, customarily equal to 24 hours.

In this case, the reference void ratio at a stress of 1 atm, e_{1v}, becomes $e_{1v}(t_{ref})$. For $t > t_{ref}$, the K_0–LCC moves proportionally downward but the slope ρ_c remains unaffected.

Accounting for viscous effects, the normalized distance can therefore be adapted by introducing a viscous coefficient ρ_α as follows:

$$\delta_b = 1 - \sigma'_v \cdot \left[e / e_{1v}\left(t_{ref}\right)\right]^{1/\rho_c} \cdot \left(t / t_{ref}\right)^{\rho_\alpha/\rho_c} \tag{5.12}$$

In a simple way, this equation allows the calculation of time-dependent behavior into the total strain.

5.3 MODEL CALIBRATION

According to Pestana and Whittle (1999), the model formulation would be strictly valid for freshly deposited soils; to be applied to the Venetian soils, it is assumed that they are characterized by very low OCRs (namely those measured at Treporti with an average OCR in the range 1.1–1.3 and not exceeding 2.5).

For these silty-clayey soils, it is very difficult to assess the value of OCR by estimating the preconsolidation stress from oedometric tests and the compression response in the transitional regime (namely, before the LCC). The transition to the LCC is therefore not necessarily related to the preconsolidation stress as is the case for high clay content materials. Although further research is required to corroborate this fact, some results indicate that for clay contents larger than 25–30%, the 'break' in the compression curve represents the preconsolidation stress, but this may be a function of clay mineralogy.

To calibrate the model, elastic and plastic parameters are needed.

Regarding the elastic component, elastic shear stiffness has been determined using Equation (5.3) with $G_b = 538$, $m = 1.1$, and $n = 0.66$, selected according to Cola and Simonini (2002) and representative of the trend for all data. Since variations in the coefficient of thrust at rest K_0 and Poisson's ratio ν have a minor effect on the predicted response, they were assumed constant and $K_0 = 0.5$ and $\nu = 0.30$ for all the Venetian materials.

Plastic response parameters have been evaluated from several oedometric tests carried out on samples collected from three boreholes located close to one another at the center of the MTS, at depths between 13 m and 94 m below mean sea level. Details can be found in Biscontin et al. (2007).

More particularly the following tests have been interpreted:

1) To compare the differences in one-dimensional response of all the natural samples within the customarily applied laboratory stress range, tests on undisturbed natural samples, belonging to the three soil classes SM-SP, ML, and CL, loaded up to a vertical effective stress in the range 6–12 MPa (Biscontin et al. 2001; Cola and Simonini 2002);
2) To evaluate the slope ρ_c in the K_0–LCC regime, tests on reconstituted samples obtained using SP-SM and ML soils from the first series, loaded up to high stresses $\sigma'_v = 34$ MPa;
3) To evaluate the influence of FF on ρ_c and e_1 tests on reconstructed mixtures of SP and CL prepared at different sand vs. clay percentages, loaded up to high stresses $\sigma'_v = 34$ MPa.

The stress level (6–12 MPa) applied in the first group of undisturbed samples was not sufficiently high to induce a well-defined LCC regime and to estimate the slope ρ_c for SM-SP and ML classes. The experimental database on natural samples was therefore integrated with the second group of oedometric tests, carried out in a very stiff piston oedometer with reconstituted SP-SM and ML specimens (prepared by pluvial deposition in the oedometer) that were loaded up to stress levels of 34 MPa, beyond the crushing regime. In addition, loose and dense specimens were tested for each soil component to evaluate the effectiveness of the testing procedure in reaching the LCC regime.

On the basis of compression curves normalized by its reference void ratio e_{1v}, the Venice soils have been divided into two groups. The first group (Group 1), including all the SM-SP and ML samples (see Figure 5.3a and b), is characterized by a K_0–LCC average slope equal to -0.24; the slope $\rho_c = 0.24$ also fitted the majority of the curves for the CL samples as shown in Figure 5.4, excluding very few specimens, composed of organic clays, which provided an average value $\rho_c = 0.40$ (Group 2), as depicted the same figure. This is a particularly interesting observation that confirms the basic idea that Venetian sediments characterized by the same mineralogical composition should have the same ρ_c.

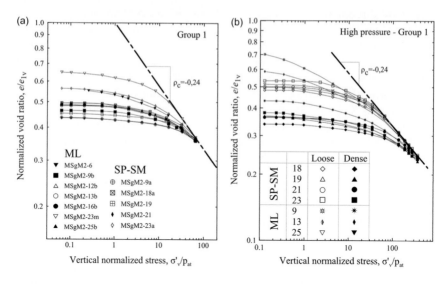

Figure 5.3 One-dimensional normalized curves of (a) natural and (b) reconstituted samples at high stress of SM-SP and ML soils (Biscontin et al. 2007).

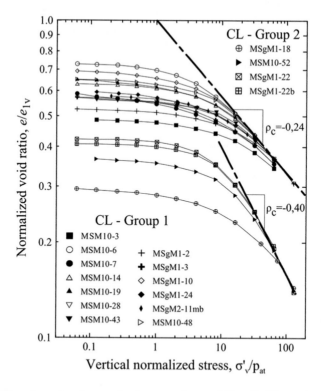

Figure 5.4 One-dimensional normalized curves of natural CL soils (Biscontin et al. 2007).

The third series of tests was carried out to properly evaluate the transition between granular and fine matrix response. The artificial specimens were prepared by mixing different amounts of sand SM-SM and silty clay CL. The silty clay, homogenized to slurry at a water content equal to 1.5· LL, was mixed with sand in percentages equal to 25%, 50%, 60%, and 70% by weight. Pure sand and pure silty clay were also tested. These sand–silty clay mixtures were loaded up to high stress 34 MPa in the oedometer apparatus. The results of these tests are the curves of Figure 5.5 normalized by its reference void ratio e_{1v}.

All the normalized curves, independent from grain size composition, approach the K_0–LCC line of Group 1, characterized by $\rho_c = 0.24$.

Figure 5.6 shows the variation of e_{1v} with FF. Experimental data at high FF corroborate that e_{1v} is uniquely dependent on FF and e_{c1v} (see Equation 5.10); the interpolation on the data is extremely good for Group 2, while the scatter is larger for Group 1. This may be related to the highly heterogeneous nature of Venetian soils, which causes gradation to vary significantly even at the centimeter scale, as shown by Cola and Simonini (1999). The results at high FF also support the existence of two different mineralogical

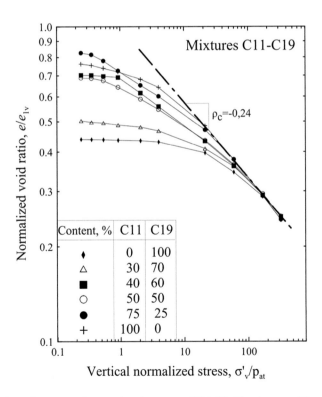

Figure 5.5 One-dimensional normalized curves of SM-SP-CL mixtures (Biscontin et al. 2007).

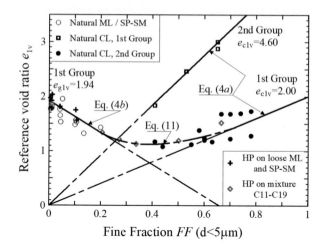

Figure 5.6 Experimental reference void ratio vs. fine fraction (Biscontin et al. 2007).

compositions for the samples, since the data align on two different lines, corresponding to different values of e_{clv}. Data regression shows a larger scatter for the coarse-grained fraction of Group 1; they may be interpolated according to Equation (5.12) with $e_{glv} = 1.94$, as shown in Figure 5.5. The reference void ratios for the third series of oedometric tests, also plotted in Figure 5.6, show a continuous trend in the range $0.20 < FF < 0.7$, confirming the gradual transition from a response governed by the granular fraction to one controlled by the water–clay matrix. Within this range, e_{lv} is affected by both e_{glv} and e_{clv}, since the coarse particles are neither in full mutual contact nor float in the clay matrix.

For materials belonging to Groups 1 and 2, the transition can be described, respectively, by the following relationships:

$$e_{lv} = e_{glv} \cdot \exp(0.25 - 4.76 \cdot FF) + (e_{clv} - 0.12) \cdot FF \qquad (5.13)$$

$$e_{lv} = e_{glv} \cdot \exp(0.15 - 6.45 \cdot FF) + (e_{clv} - 0.20) \cdot FF \qquad (5.14)$$

These relationships (Equations (5.13) and (5.14)) are valid for FF in the range between 0.20 and 0.70 (Group 1) and FF between 0.13 and 0.50 (Group 2), respectively.

The values of parameter θ, affected by grain size distribution and the shape of grains, have been determined for Venice soils from the best fitting of one-dimensional compression curves of the first series of oedometric tests and plotted in Figure 5.7 as a function of the uniformity coefficient, U. For comparison, data from sands and sandy silts reported by Pestana (1994) are also reported in the same figure.

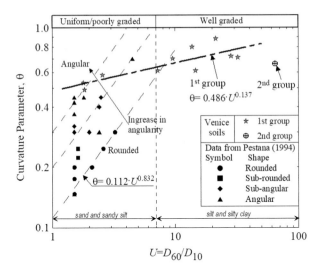

Figure 5.7 Relationship between model parameter θ and non-uniformity coefficient U (Biscontin et al. 2007).

Considering the sub-angular and angular shapes of the poorly graded Venice soils, θ lies at the upper boundary of the existing data and is slightly dependent on the non-uniformity coefficient U through the following equation:

$$\theta = 0.486 \cdot U^{0.137} \ \left(r^2 = 0.68\right) \tag{5.15}$$

The reliability of the proposed model to describe the one-dimensional compression response of Venetian soils and the oedometric curves of undisturbed specimens from the three classes SM-SP, ML, and CL (the latter belonging to both Groups 1 and 2) were calculated using the model parameters selected according to the calibration procedure described above.

From the curves plotted in Figure 5.8a–c, a good agreement between measured and predicted compression behavior can be noticed, showing the nonlinear compression curves in the transition regime approaching the K_0–LCC in the log$e$ – logσ'_v plane at higher stresses.

It is interesting to note that thanks to the above model parameter calibration, it would theoretically be possible to calculate the settlement induced by vertical loading in the entire lagoon area through the constitutive model and on the basis of grain size composition and in situ void ratio.

Another important point to consider is that model predictions are very sensitive to the in situ void ratio, which can rarely be defined with accuracy. The data dispersion of Figure 5.6 can be attributed to this difficulty. Therefore, the simulation of compression curves proposed in Figure 5.8a–c

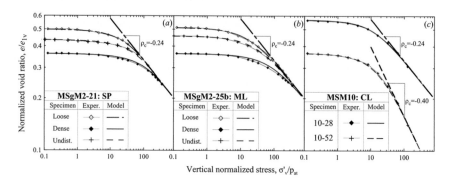

Figure 5.8 Comparison of experimental vs. modeled results for SP (a), ML (b), and two CL (c) soils (Biscontin et al. 2007).

were obtained using the values of parameter e_{1v} measured in each experimental test, rather than predicted by Equations (5.10 and 5.11) and (5.13 and 5.14).

In addition, the transition between lower and higher stress response occurs gradually in the experimental curves without any yielding point (as is common in most elastoplastic models), either for coarsely grained or for finely grained soils. The OC state is consequently considered only for its effect on the value of the initial void ratio, considering all the curves as first-loading curves.

In the next section, however, it will be shown how the additional effect of introducing a yielding stress can be used to separate the elastic response from the elastoplastic one and to more properly predict actual in situ soil behavior.

5.4 BACK-ANALYSIS OF SETTLEMENTS OCCURRED AT THE *TREPORTI TEST SITE*

To evaluate the applicability of the compression model to calculate settlements in the Venetian area, the results from the Treporti test site described in Chapter 4 have been used.

Figure 5.9 sketches the local strain against depth provided by the sliding deformeter (SD) n. 3, at increasing height of the bank; for more careful checking of the local stratigraphic profile, the CPTU trend is also reported.

It is assumed that below the center of the embankment a nearly one-dimensional compression condition occurs and that the increments of stress induced by the applied load can be determined according to the solution provided by Poulos and Davis (1974) valid for a circular loaded area over a linearly elastic homogeneous isotropic half-space. The calculation was carried out down to 30 m below the ground surface.

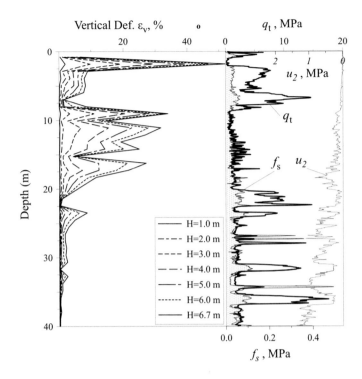

Figure 5.9 Evolution of settlement during embankment construction.

The soil profile was divided into 1 m thick layers; on the basis of the grain size distribution and of the Atterberg limits (fundamental to attribute each sublayer to Group 1 or 2), each 1 m thick layer was then subdivided into homogeneous sublayers and for each of them the material parameters required by the constitutive model were selected and assigned. CPTU profiles were used to control the validity of the sub-layering subdivision.

Figure 5.10 shows the values of selected ρ_c, FF, e_0, and U from which all the model parameters were determined.

The strains were calculated for any sublayer and then integrated over each 1 m thick layer, to get the local strain to be compared with the measurements provided by the central SD n.3.

The model was applied to two different hypotheses:

a) The basic formulation of the model was assumed, that is, elastic and plastic strains occur from the beginning of the loading phase, with no time effect that is $t/t_{ref} = 1$;

b) The existence of a yielding stress (related to OCR) was hypothesized, below which the behavior is elastic and beyond which it is elastoviscoplastic. In this case the elastic shear modulus was reduced from

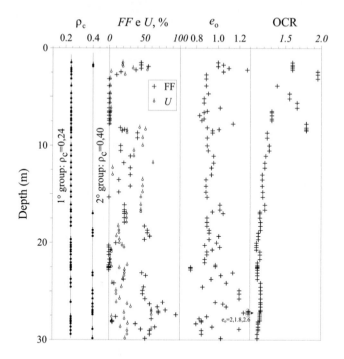

Figure 5.10 Profile of the model parameters.

the maximum stiffness G_0 to an operative value equal to 0.25 G_0. The viscous behavior was added according to Equation (5.12), by assuming the real timing of each construction step until embankment completion.

The introduction of a yielding stress was suggested by the observation that the field response at small stress levels is stiffer than that observed in the laboratory in the oedometric tests, for all the Venetian soils. This aspect was obvious at the *Treporti test site*, whose results in terms of vertical strain vs. vertical stress allowed for the estimation of a yielding stress (assumed as preconsolidation stress) for all the soils, including sands. The OCR turned out to be between 1.5 and 2.5 for the shallower Holocenic soils and less than 1.3 for the Pleistocenic sediments, as shown in Chapter 4.

This overconsolidation state may be prevalently attributed to field aging over millennia and centuries after the deposition.

Venetian sediments in the lagoon area dated to around 15,000 years at a depth of 20 m below ground level (Bonatti 1968). A tentative continuous OCR profile was determined from the following relationship (Ladd 1971):

$$OCR = (t/t_0)^{[C_\alpha/(C_c-C_r)]}$$

(5.16)

in which t_0 = a reference time (assumed equal to 1 year) and C_c, C_r, and C_α are the primary, unloading–reloading, and secondary compression indexes. Assuming a $C_\alpha /(C_c - C_r)$ = 0.028 and considering the sediment age, a value of OCR = 1.28 is obtained, which is consistent with in situ measured values.

Figure 5.10, last column on the right, depicts the profile of OCR used in the constitutive model application.

Regarding reference time used in the model simulation to accomplish case (c), t_{ref} = 2 hours was assumed for all the Venetian soils accounting for their high drainage properties. According to Pestana and Whittle (1998), parameter ρ_α was set to 0.006 and 0.010 for Groups 1 and 2, respectively, hypothesizing $\rho_\alpha /\rho_c = C_\alpha /C_c$ 0.025 for both the soils over the entire compression range.

Figure 5.11 compares the field compression curves measured at TTS and the model predictions for three layers (3.4–4.4 m, 10.6–11.6 m, and 23.0–24.0 m below g.l.) composed of different classes of soil, namely medium-fine uniform sand, clayey silt, and alternance of silty and organic clay.

Figure 5.12 displays the local settlement vs. depth corresponding to half bank (H = 3.5 m) and completed bank (H = 6.7 m) with the assumptions of cases (a) and (b).

Note the satisfactory prediction, especially for the shallowest 20 m, of the constitutive model at bank completion for each layer, even using the simplest formulation, with no time and overconsolidation effects. The calculated total settlement is equal to 366 mm, whereas the measured one is 325 mm, which is an overestimation of about 12%.

This result is extremely satisfactory and demonstrates that the constitutive model is able to predict settlements over sand, silts, and clay in complex heterogeneous soil conditions typical of the Venice lagoon, exclusively on the basis of grain size distribution and in situ void ratio.

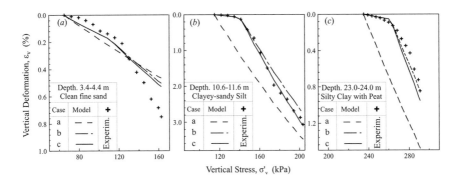

Figure 5.11 Comparison of experimental vs. modeled results of three different layers.

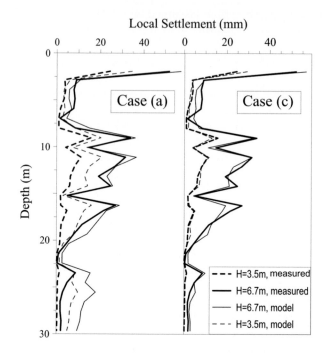

Figure 5.12. Measured vs. calculated settlements, cases (a) and (c).

From a general point of view, the model tends to overestimate strain in clayey soils and underestimate strain in sandy materials. This appears to be due to principally the lack of precise knowledge of the parameter in situ void ratio e_0 with depth, which affects the expected result more than others.

A major difference is observed for a bank height of 3.5 m; in this case, the effects of overconsolidation with a stiffer response in the field is pronounced, showing that the model predicting elastoplastic strain since the beginning of the loading phase may lead to an overestimate of settlements, especially in finely grained formations. By introducing the yielding stress and time effect, the model seems to be able (see Figure 5.12b) to accurately predict the settlement in both construction phases. However, it must be pointed out that determining the OCR profile is one of the major issues in Venetian soils.

The above observations demonstrate that a unique and simple set of constitutive equations are able to provide an excellent tool for calculating settlement over highly heterogeneous silty-based soils. One of the major difficulties encountered in application of the model is ultimately overcome by knowledge of a detailed estimate of the initial void ratio for the different sublayers, which should be subdivided, and this can significantly affect the predictable subsoil response.

REFERENCES

Biscontin, G., Cola, S., Pestana, J. M. and Simonini, P. 2007. A unified compression model for the Venice lagoon natural silts. *Journal of Geotechnical and Geoenvironmental Engineering* 133(8), 932–942.

Biscontin, G., Pestana, J. M., Cola, S. and Simonini, P. 2001. Influence of grain size on the compressibility of Venice Lagoon soils. *Proceedings of the XV ICSMFE*, Balkema, Rotterdam, The Netherlands 1, 35–38.

Bonatti, E. 1968. Late-Pleistocene and postglacial stratigraphy of a sediment core from the Lagoon of Venice (Italy). *Mem. Biog. Adr., Venezia* VII(Suppl.), 9–26.

Cola, S. and Simonini, P. 1999. Some remarks on the behaviour of Venetian silts. *Proceedings 2nd International Symposium on Pre-Failure Deformation Characteristics of Geomaterials*, Balkema, Rotterdam, The Netherlands, 167–174.

Cola, S. and Simonini, P. 2002. Mechanical behaviour of the silty soils of the Venice lagoon as a function of their grading characteristics. *Canedian Geotechnical Journal* 39, 879–893.

Cortellazzo, G., Ricceri, G. and Simonini, P. 2007. Analisi retrospettiva del comportamento del terreno di fondazione di un rilevato sperimentale ubicato nella laguna di Venezia. *Proceedings of XXIII Convegno Nazionale di Geotecnica*, 16–18 Maggio 2007, Padova, Patron Editore. (*in Italian*).

Ladd, C. C. 1971. *Strength Parameters and Stress-Strain Behaviour of Saturated Clays*. Res, Rep. R71–23, Cambridge: MIT.

Li, X. S. and Dafalias, Y. F. 2000. Dilatancy for cohesionless soils. *Géotechnique* 50(4), 449–460.

Mitchell, J. K. 1976. *Fundamentals of Soil Behavior*. New York: Wiley.

Pestana, J. M. 1994. *A Unified Constitutive Model for Clays and Sands*. ScD Dissertation, Massachusetts Institute of Technology, Cambridge.

Pestana, J. M. and Whittle, A. J. 1995. Compression model for cohesionless soils. *Geotechnique* 45(4), 611–632.

Pestana, J. M. and Whittle, A. J. 1998. Time effects in the compression of sands. *Géotechnique* 48(5), 695–701.

Pestana, J. M. and Whittle, A. J. 1999. Formulation of a unified constitutive model for clays and sands. *International Journal for Numerical and Analytical Methods in Geomechanics* 23(12), 1215–1243.

Poulos, H. G. and Davis, E. H. 1974. *Elastic Solutions for Soil and Rock Mechanics*. London: John Wiley & Sons.

Rocchi, G., Fontana, M. and Da Prat, M. (2003). Modelling of natural soft clay destruction process using viscoplasticity theory. *Géotechnique* 53(8), 729–745.

Chapter 6

Geotechnical behavior
of defense structures

Written with Giorgia Dalla Santa

6.1 INTRODUCTION

As already stated in previous chapters, MoSE is an integrated plan of interventions implemented across the entire lagoon area to safeguard Venice and its lagoon from high tides. The MoSE mobile barrier system, almost completed in 2020 and now operative in case of predicted very high tides is the central element and the most challenging part of the interventions. MoSE is an acronym that stands for 'Modulo Sperimentale Elettromeccanico'.

The technical solution adopted combines design constraints and legislative requirements defined by the central government in terms of environmental impact and minimum interference with local economic activities.

It basically consists of four barriers, as shown in Figure 6.1.

Since the Lido inlet is 1.5 km wide, it was decided to subdivide the inlet into two parts by building an artificial island made of sand in the middle of the inlet. The island splits the barriers into two sections known as the Treporti barrier (with 21 gates, for a total length of 420 m) and the San Nicolò barrier (made up of 20 gates, for a total length of 400 m). At the Malamocco and Chioggia inlets, there are single sets of barriers with 19 gates (total length of 380 m) and 18 gates (total length of 360 m), respectively. The dimensions of the gates and the foundation caissons vary from barrier to barrier according to the depth and width of each inlet.

The MoSE system consists of fold-away steel gates hinged to large concrete caissons buried in the subsoil at the inlets. The gates are generally filled with water, folded and embodied in the caissons (Figure 6.2, upper part). When tides exceed +1.1 m, the gates are lifted by pumping in the air (Fig. 6.2, lower part) to expel the water, thus isolating the lagoon from the northern Adriatic Sea for the amount of time necessary to attain a drop in the sea level. It customarily takes 30 min to elevate each gate, whereas the time needed to fold the gates by expelling the air is 15 min. The gates can sustain a maximum difference of the hydraulic head between the sea and the lagoon of around 2.0 m.

Furthermore, each inlet is defended by complementary structures represented by outside breakwaters made with rip-rap and accropodes with

DOI: 10.1201/9781003195313-6

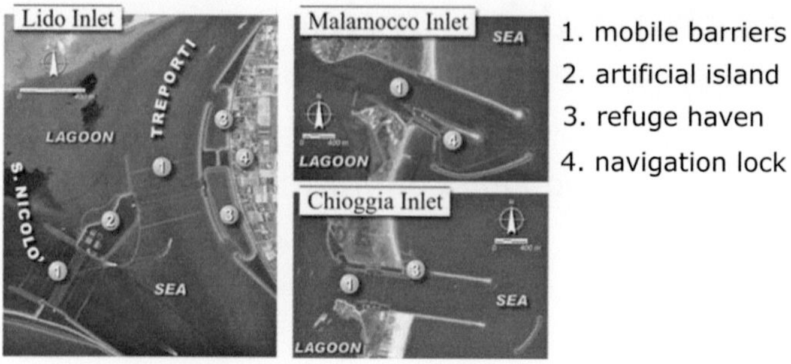

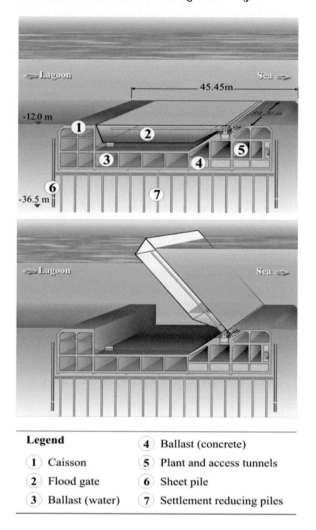

Figure 6.1 Location of the mobile barriers at the lagoon inlets (Jamiolkowski et al. 2009).

Legend		
	(4) Ballast (concrete)	
(1) Caisson	(5) Plant and access tunnels	
(2) Flood gate	(6) Sheet pile	
(3) Ballast (water)	(7) Settlement reducing piles	

Figure 6.2 Gates embodied in the caissons and in action (Jamiolkowski et al. 2009).

lengths varying from 1,280 m (at the Malamocco inlet) to 520 m (at the Chioggia inlet).

6.2 GEOTECHNICAL INVESTIGATIONS ALONG THE BARRIER SECTIONS

Geotechnical investigations were performed in the late 1980s to draw relevant soil profiles along with cross sections of the three lagoon inlets and to estimate classical geotechnical properties for a preliminary selection of the barrier foundations. For a proper final foundation design, all the previous investigations were integrated, between 2001 and 2004, with additional and more exhaustive site and laboratory testing, for example:

- 88 boreholes with undisturbed 98 mm Osterberg sampling, at depths from 40 m to 120 m below the sea floor;
- 13 standard penetration test (SPT) dedicated boreholes, depth 40–50 m;
- 119 CPTU, depth 30–120 m;
- 11 DMT, depth 50 m;
- 3 crosshole seismic tests (CHT), depth 80 m.

Accounting for the results of MTS and TTS and for all the geotechnical investigations carried out over the years, including those used for the final design, the subsoil profiles have been drawn along the barrier axes, as shown in Figure 6.3a, b, and c. The displayed corrected piezocone tip resistance profiles prove once more the pronounced spatial variability of the lagoon deposits at the three inlets.

Note that the bottom of the caissons reaches depths from –24.2 m to –27.2 m from the mean sea level, approximately reaching the heterogeneous layer of clayey silt and sandy silt (layer D).

Given that the barriers at the three inlets are extremely similar, the following paragraph provides a detailed description of the geotechnical aspects of the San Nicolò barrier.

6.3 MAIN GEOTECHNICAL ASPECTS RELATED TO FOUNDATION CAISSONS FABRICATION SITES AND INSTALLATION

The design of the barrier caisson foundations at the three inlets posed many geotechnical challenges. The barriers with their operating tolerances required soil improvement to minimize settlements under the application of operating loads, which must be compatible with the deformability of the hydraulically sealed joints that are located between the caissons.

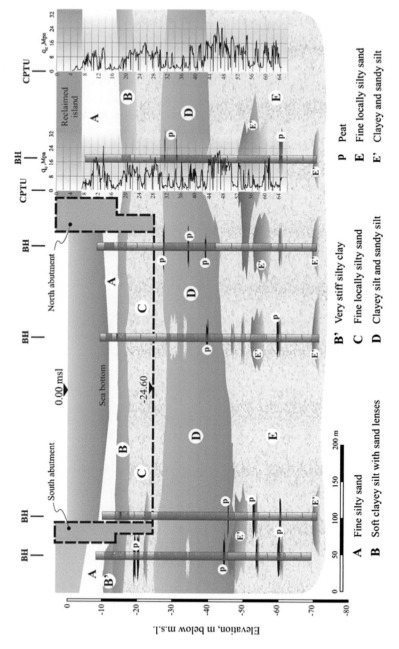

Figure 6.3 Soil profile at (a) Lido San Nicolò barrier, (b) Malamocco barrier, and (c) Chioggia barrier (Jamiolkowski et al. 2009).

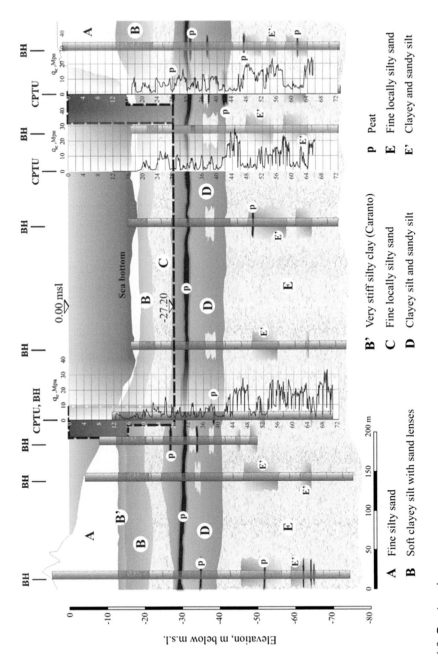

Figure 6.3 Continued.

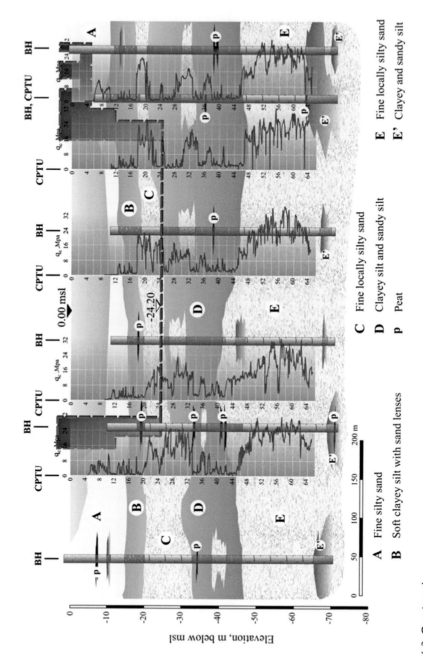

Figure 6.3 Continued.

According to Burland (1990) and Fioravante et al. (2008), settlement reducing piles were selected as one possible solution. Under the San Nicolò and Malamocco barriers, spun driven concrete piles were installed, while jet grouting columns were used at the Treporti barrier and driven steel piles at the Chioggia inlet.

One interesting aspect of the construction regards the two different ways in which the foundation caissons were manufactured. Part of the foundation structures was fabricated in a 400 × 800 m temporary embankment (+2 m from the medium sea level), built expressly for this purpose at the Malamocco inlet. The reinforced concrete caissons, whose weight is about 245200 kN were built on this embankment over hundreds of pillars, in order to allow specifically built trains of coordinated hydraulic jacks to go underneath, lift the caissons, and finally, after running along special rails, bring the caissons to a special mobile platform called 'syncrolift' (see Figure 6.4). The required perfect alignment of the rails imposed an extremely low settlement tolerance. At this point, each foundation caisson was then lowered into the water by the *syncrolift* which worked like a sort of giant elevator, equipped with pulleys. The caissons were then floated on the water and dragged to their final position where they were sunk.

The remaining foundation caissons were fabricated on two dry docks of about 500 × 100 m, realized at the Treporti and Chioggia inlets (see Figure 6.5). To isolate the construction sites from the seawater the hydraulic belting was realized with sheet piling cofferdams and cutter soil mixing impermeable diaphragm walls down to 25 m to intercept the low permeability silty clay layer at depths between –12 m and –17 m, positioned among two sandy aquifers. Afterward, the pumping and monitoring well system was created inside and outside the belting to dewater the internal basin and carefully monitor the hydraulic conditions both of the higher and lower aquifers, both inside and outside the belting, also to maintain the previous water table condition in the outer residential area. Excavation and re-profiling activities completed the realization of the dry dock where, at the depth of –12 m, the caissons were built. The pumping and monitoring well system guaranteed the control of the underpressures and the maintenance of the bearing capacity of the dry dock bottom. At the end of the building phase, the construction sites were flooded again, so that the caissons would float and then could be guided to their final position, where they were sunk.

The caisson installation sequence, depicted in Figure 6.6 for the San Nicolò barrier, was executed as follows:

- Dredging down to 14.0 m below m.s.l.;
- Driving two alignments of sheet piles spaced 49.0 m, down to 36.5 m below m.s.l.;
- Dredging a trench between the sheet pile walls down to the foundation plane at 24. 2 m below m.s.l.;

Figure 6.4 An overview of the embankment built as foundation caissons fabrication site at Malamocco inlet. Source: Giorgio Marcoaldi-CVN; Copyright: Copyright CEMultimedia 2014 (from https://www.MoSEvenezia.eu/).

- Driving piles down to 43.6 m below m.s.l.;
- Placement of a granular fill with a thickness of 1.0 m and above it a 0.3 m thick layer of self-leveling mortar;
- Installing each caisson over the fill using four temporary saddles
- Leveling the caisson by means of hydraulic jack;
- Grouting the space between the bottom of the caisson and the granular fill.

Figure 6.5 An overview of the flooding of the dry dock, after the fabrication of the foundation caissons at Chioggia inlet in March 2014. Copyright: Copyright CEMultimedia 2014 (from https://www.MoSEvenezia.eu/).

The number of caissons in the four barriers ranges between seven and nine, each caisson housing two to three flap gates. The following are the typical caisson dimensions: 40–60 m long; 30–50 m wide; 10–12 m high.

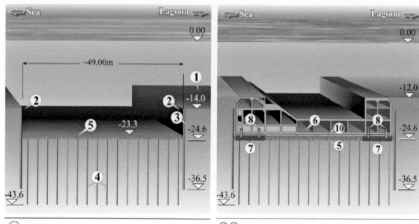

① Dredging to elev. -14.0 b.m.s.l.
② Sheet pile driving
③ Trench dredging to elev. -24.6 b.m.s.l.
④ Piles driving
⑤ Placement of granular fill 1.0 m,
   self levelling mortar 0.3 m

⑥⑦ Caisson placement on four temporary saddles
⑧ Levelling the caisson by means of hydraulic jacks
⑨ Caisson's ballasting
⑩ Grouting the space between caisson's bottom
   and granular fill

Figure 6.6 Construction sequence (Jamiolkowski et al. 2009).

Figure 6.7 shows a view of the San Nicolò barrier, which is composed of seven barrier caissons (B_1–B_7) and two abutment caissons (A_1 and A_2), all of them separated by rubber joints. These joints (J_1–J_8) must ensure waterproof contacts between adjacent caissons, thus guaranteeing accessibility to the service tunnels for maintenance of the electromechanical devices that move the flap gates.

The joint waterproofing requires strict limitations of the maximum differential settlements between two contiguous caissons defined in Figure 6.8a–c (with the sign assumption).

More particularly, Δ_1 = differential vertical displacement of two adjacent caissons must be less than ±40 mm for all the joints, Δ_2 = the rotation of

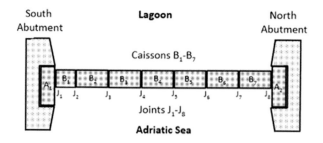

Figure 6.7 Plan view of the San Nicolò barrier.

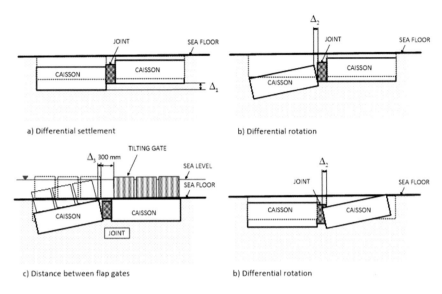

a) Differential settlement

b) Differential rotation

c) Distance between flap gates

b) Differential rotation

Figure 6.8 Definition of differential settlement and rotation. (a) differential settlement; (b) differential rotation; (c) distance between flap gates; (d) another type of differential rotation.

two contiguous caissons causing the compression or decompression of the joints must be less than +(29–40) mm (Figure 6.6b), and Δ_3 = change in the distance between the tilting gates hinged to two adjacent caissons must be +(124–127) mm (see Figure 6.8c), the latter two depending on the type of joint (Jamiolkowski et al. 2009).

The geotechnical design company 'Studio Geotecnico Italiano' selected the foundation solutions, similar for all four barriers, and carried out the calculations, which are briefly described in Jamiolkowski et al. (2009).

The long-term prediction, up to 100 years, included total and differential settlements under complex loadings accounting for self-weight, tides, and wave action as well as for low seismicity (with a peak ground acceleration PGA < 0.06 g) earthquakes.

Due to the complex construction sequence and the relevant spatial soil heterogeneity, a simplified approach to the settlement evaluation was selected, consisting of the following factors:

- Specific soil profiles under each caisson were sketched and used for the calculations;
- The construction sequence, together with granular fill and subsequent caisson placement, was simulated;
- The stress increments in the ground below the caisson were computed using the linear elasticity theory;

- The initial conditions to compute settlements in the center and at the caisson's four corners were assessed based on the stress increments;
- The settlement trend of fine-grained soils over time was estimated by a viscous-plastic coupled consolidation soil model (Zeevaert 1972; Rocchi et al. 2003), adapted to the one-dimensional compression;
- The settlement of coarse-grained soils in drained condition was computed using the elastoplastic strain hardening model proposed by Li and Dafalias (2000).

All the material model parameters were estimated from laboratory tests, seismic in situ tests, as well as the trial embankment built up at the TTS.

The stiffness of the soil, treated with settlement reducing piles, was determined using the homogenization approach. Under all the caissons of the San Nicolò barrier, the equivalent stiffness varies in a wide range between a minimum of 32 MPa under caisson n. 2 to a maximum of 82 MPa under the North Abutment caisson.

6.4 SETTLEMENTS UNDER THE BARRIERS

The settlements, calculated along with the construction sequence over a time of 100 years, showed relevant differential displacements and rotations between caisson n. 1, n. 2, and n. 9 alongside the abutments. These are due to the load applied by the reclaimed sand-fill placed behind the abutment caissons and dependent on the construction sequence over time accounted in the calculation.

The computed maximum differential displacements, listed in Table 6.1, are generally below the acceptable limits with the exception of the two abutment caissons and the barrier caissons n. 6 close to the abutments. The larger differential settlements at these spots are caused by the placement of reclaiming fill behind the abutment caissons and by the construction of the artificial island that splits the Lido inlet into two parts.

As discussed above, the concrete caissons in the San Nicolò barrier are installed over spun and driven concrete piles, with an overall length of 19 m and a diameter of 500 mm. These have been driven after dredging between sheet piles (down to the sea floor, at –24.6 m below m.s.l.). The pile heads have been covered by 1 m thick compacted granular fill and 300 mm of self-leveling mortar to obtain a suitable interaction between the improved ground and the caisson slab.

This improvement had already been proposed in the preliminary design stage (early 1980s), to reduce the differential settlement due to the spatial soil heterogeneity and any possible construction imperfection related to the difficulties of the works below the sea floor.

Table 6.1 Maximum computed relative displacements

Joint	D_1 (mm)	D_2 (mm)	D_3 (mm)
J_1	17 (30y)	<5	7 (100y)
J_2	55 (100y)	<5	<5
J_3	−15 (100y)	<5	<5
J_4	−8 (EOC)	<5	<5
J_5	11 (100y)	<5	<5
J_6	17 (100y)	<5	<5
J_7	−9 (EOC)	<5	7 (EOC)
J_8	−19 (30y)	15 (100y)	51 (100y)

EOC = end of construction.

The interaction between the driven piles and the surrounding soil was modeled using the code NAPRA (Russo 1998; Mandolini et al. 2005), whose principal features are listed below:

- Stratified linear elastic half-space;
- Variable raft stiffness;
- Variable piles geometry and material properties;
- Tensionless raft–soil contact;
- Interaction at the raft–pile contact, nonlinear elastic;
- Interaction among piles, linear elastic, assuming a cut-off at the certain distance as suggested by Randolph and Wroth (1978);
- Piles connected to the raft by hinges.

The results of the simulations provided by NAPRA for one of the standard caissons belonging to the San Nicolò barrier are depicted in Figure 6.9, which reports two basic pieces of information for the design of rigid rafts on settlement reducing piles:

α_p = vertical load on piles/total vertical load,
α_s = settlement of piled foundation/settlement of unpiled foundation.

Taking the limitation on differential settlements and rotations between the caissons into account, it was decided to select a spacing = 3.8 m among the driven piles. With such spacing, the total acting permanent load of 158,626 kN corresponding to the total number of 268 piles gives $\alpha_p = 0.72$ and $\alpha_s = 0.61$. The loads acting on the piles through the rafts in different conditions of the tilting gates are listed in Table 6.2.

The results obtained by the simplified NAPRA code have been validated by means of a comprehensive series of centrifuge tests carried out

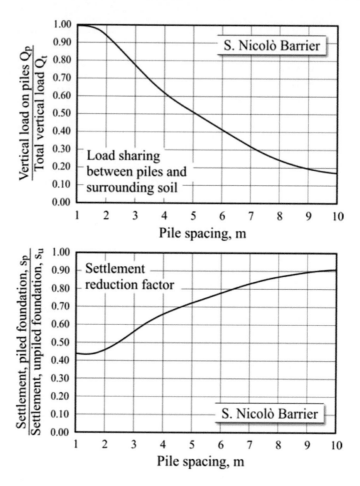

Figure 6.9 Results of the analyses carried out with the code NAPRA (Jamiolkowski et al. 2009).

at ISMGEO (Fioravante et al. 2008) and by finite element analyses, performed using the same material models used for the settlement evaluation (Jamiolkowski et al. 2009).

Physical and numerical modeling confirmed the response of NAPRA code in terms of both load sharing mechanism (α_p) and settlement reduction factor (α_s).

Moreover, it has been shown that the presence of a granular layer between the caisson bottom and the pile heads produces negative shaft friction along almost the entire upper half of the piles, as shown in Figure 6.10.

This aspect, which does not particularly affect the overall response of the caisson–piles–subsoil, can be explained by considering that the insertion of

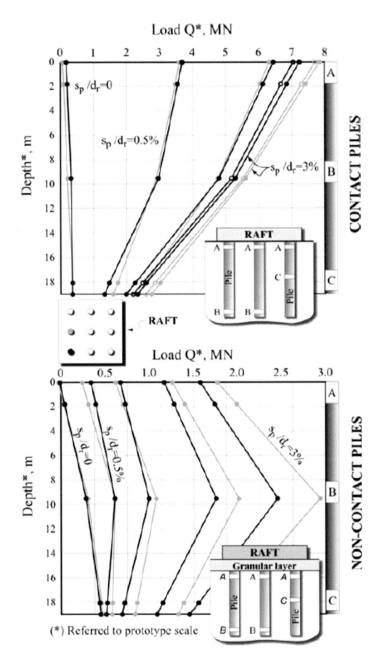

Figure 6.10 Centrifuge experiments on the raft on piles without (a) and with (b) granular layer (Jamiolkowski et al. 2009).

Table 6.2 Loads on piles

Loading conditions	Q_{pmax} (kN)	Q_{pmin} (kN)	Q_{pave} (kN)	$\Sigma Q_p/Q_t$ (%)
Gate quiescent	725	535	614	66
Gate in action	793	520	642	65

Q_p = load on a pile, Q_t = total vertical load.

a deformable layer between the caisson slab and the pile head inhibits the effective displacement compatibility between the raft and the piles.

After installation, a monthly topographic survey through the tunnel connecting all the caissons was carried out.

From a general point of view, the measured settlements are very small, a bit less than the predicted ones, suggesting the effectiveness of the solution adopted to improve subsoil stiffness below the barrier caissons at San Niccolò barrier and at the other two barriers, Lido and Malamocco, which adopted similar soil improvement methods. Slightly higher settlements have been measured at the abutments as a consequence of higher loads transferred by the caissons to the subsoil.

REFERENCES

Burland, J. B. 1990. Piles as settlement reducers. *Proceedings of XIX Convegno Nazionale di Geotecnica, A.G.I., Pavia* 2, 21–24. (*in Italian*).

Fioravante, V., Giretti, D. and Jamiolkowski, M. 2008. Physical modelling of raft on settlement reducing piles, *Symposium Honoring Dr. John H. Schmertmann for His Contributions to Civil Engineering*. In *From Research to Practice in Geotechnical Engineering*, 206–229.

Jamiolkowski, M., Ricceri, G. and Simonini, P. 2009. Safeguarding Venice from high tides: Site characterization and geotechnical problems. Keynote lecture. In *Proceedings of the 17th International Conference on Soil Mechanics and Geotechnical Engineering* (Volumes 1, 2, 3 and 4). October 4–9. Alexandria: IOS Press, 3209–3227. From the pre-print version available at UNIPD IRIS website https://www.research.unipd.it/item/preview.htm?uuid=4b48dac1-44ac-4dd9-9e4f-9b11b7dc7395.

Li, X. S. and Dafalias, Y. F. 2000. Dilatancy for cohesionless soils. *Géotechnique* 50(4), 449–460.

Mandolini, A., Russo, G. and Viggiani, C. 2005. Pile foundations: Experimental investigation, analysis and design. In *Proceedings of the 16th International Conference on Soil Mechanics and Geotechnical Engineering* (Vol. 16, No. 1). Millpress Science Publisher/IOS Press. Doi: 10.3233/978-1-61499-656-9-177.

MoSE Venezia – official site. https://www.MoSEvenezia.eu/.

Rocchi, G., Fontana, M. and Da Prat, M. 2003. Modelling of natural soft clay destruction process using viscoplasticity theory. *Géotechnique* 53(8), 729–745.

Russo, G. 1998. Numerical analysis of piled rafts. *International Journal of Numerical and Analytical Methods in Geomechanics* 22(6), 477–493.

Zeevaert, L. 1972. *Foundation Engineering for Difficult Subsoil Conditions.* New York: Van Nostrand Reinhold Company.

Chapter 7

Geotechnical monitoring of marshes and wetlands

Written with Giorgia Dalla Santa

7.1 INTRODUCTION

As previously stated, the city of Venice lies in an extremely precious environment, that is, the Venice lagoon. As is typical of tidal environments characterized by low hydraulic energy in coastal plains slightly tilting toward the sea, the landscape typifying the Venice lagoon is largely made up of salt marshes and wetlands, which are covered for the most part by halophytic vegetation, i.e., vegetation adapted to a salty environment, and subject to tidal fluctuation, with periods of submersion depending on tidal cycle amplitude and local topography. The importance of coastal salt marshes derives from several positive effects they have on the ecosystem; marshes dissipate waves and reduce wind impact and erosion during storms, filter nutrients and pollutants from the water, and constitute the proper ecosystem for coastal communities of birds and fish (Tommasini et al. 2019). Marshes represent the fundamental element of the lagoonal ecosystem and maintain an equilibrium with continuous tidal events, variable water salinity (from freshwater to salty sea water), sediment erosion, and deposition. Figure 7.1 presents some views of this highly complex ecosystem of natural salt marshes and wetlands.

Since it is constantly exposed to variable driving forces, the lagoon environment and its components are in continuous transformation. Salt marsh growth, maintenance, and/or disappearance are governed by a variety of physical and biological processes (Perillo et al. 2018). Tidal marshes can evolve both horizontally, because of lateral erosion mainly due to wave and wind erosion and progradation processes, and vertically, as a net result of organic and inorganic deposition, surface erosion, and relative sea level rise (Tommasini et al. 2019). The main external controls are the sea level, tidal and sediment-supply regimes, from both the sea and the rivers flowing into the lagoon. During submersion periods, at maximum high tide, the suspended sediments deposit, resulting in an upward accretion. Conversely, during low tides, the runoff along the tidal channels crossing the marshes can erode and carry away the sediments. Intrinsic influences are provided by halophytic vegetation that produces organic matter deposition and

DOI: 10.1201/9781003195313-7

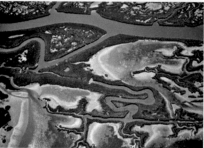

Figure 7.1. The highly complex ecosystem of natural salt marshes and wetlands (from Consorzio Venezia Nuova archive).

provides sediment autocompaction, as well as higher resistance to erosion (Allen 2000).

7.2 SEDIMENT BALANCE, SALT MARSH EROSION, PROTECTION, AND RESTORATION

The delicate 'altimetric' equilibrium of the salt marsh environment, and its capacity to grow along with high tides and submersion periods, is strictly linked to the availability of sediments. In recent decades, an erosion trend has occurred in the Venice lagoon. Erosion degrades wetlands, removes tidal flats, and flattens the lagoon bed, thus eliminating morphological elements typical to the lagoon environment, in favor of the more simplified and undifferentiated features of the marine environment. Lagoon erosion is a detriment to water circulation and biodiversity, reducing the diversification typical of the lagoon environment. The most evident consequences are the progressive silting up of channels and canals and a drastic reduction in the extent of the salt marshes, which have decreased by about two-thirds since the beginning of the 1800s, with consequent loss of habitats for plants and wildlife (Guerzoni and Tagliapietra 2006; Carniello et al., 2009).

The erosion trend is caused by the interaction of several factors, both natural and anthropogenic. First of all, the alteration of the sediment transport balance due to the lower intake of sediments caused by the diversion of several rivers that previously flowed into the lagoon (the Brenta, Piave, and Sile rivers), was carried out in the past centuries by the Republic of Venice in order to safeguard the navigation of the inner lagoon, as already introduced in Chapter 2 (Amos et al. 2010). The erosion tendency is also due to the invasive series of interventions carried out on the mountainside of the river basins, which has caused depletion of the intakes coming from the sea through the coastal current and through the inlets. Erosion has also been caused by stabilization of the inlets' morphology for safe navigation, with the construction of jetties (those at the Malamocco inlet were built between 1808 and 1840, at the Lido inlet between 1890 and 1910, and at the Chioggia inlet between 1911 and 1933). Construction of the jetties changed the inlets' hydrodynamics by lowering the inlet channel bottom, reinforcing the current velocity, and finally enhancing erosion and drop out of the sediments. The jetties have interrupted the sand transport patterns along the shore and, consequently, the sediment intake from the sea (Tambroni et al. 2005; Helsby et al. 2006). Second, the year-by-year increase in wave motion produced by wind and motorboats, which are ever-increasing in frequency and velocity, erodes the salt marsh borders (Fagherazzi et al. 2007). Finally, the relative sea level rise resulting from the interaction of eustacy and land subsidence counteracts the drawing of material from offshore to compensate for losses (Tosi et al. 2009; Amos et al. 2010).

Thus, as already mentioned in Chapter 2, a constant reduction in areas occupied by salt marshes has been registered from the comparison of historical maps. Fletcher et al. (2004) consider the construction of fish farms in the southern inner lagoon to have induced the most significant loss (20%). Guerzoni and Tagliapietra (2006) estimated from the analysis of historical maps that there were 168 km² of salt marshes in 1930, 105 km² in 1970, and 60 km² in 2002 (see Figure 7.2). The role of tidal inlets on the morphological evolution of the lagoon and the complex interactions among tidal channel hydromorphology, marsh development, inlet hydrodynamics, and sediment balance have been investigated through numerical modeling (Umgiesser et al. 2006; D'Alpaos et al. 2006; Carniello et al. 2009; Perillo et al. 2018).

In order to counteract this tendency, different protective measures have been developed, such as tidal flat reconstruction or edge reinforcements to protect natural salt marshes from erosion. The applied technique can

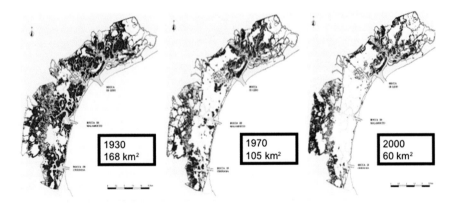

Figure 7.2. Evaluation of the diminution in the area of salt marshes (modified from Guerzoni and Tagliapietra (2006).

be diversified according to water depth and the degree to wave exposure, by applying removable modular mattresses made with various types of fibers and filled with natural materials characterized by varying degrees of resistance and durability (for example, shells, rattans, and canes) or made according to other naturalistic engineering techniques at a low environmental impact (see Fig. 7.3a).

In addition, in recent decades several salt marshes and tidal flats have been reconstructed in their original location, by using a part of the sediments obtained from dredging the navigation lagoon channels (2.3 Mm³/annum), despite the fact that a large portion of the dredges could not be used due to contamination (1.5 Mm³). First, an edge reinforcement made of wood and geosynthetic material is laid to define the shape of the new salt marshes and to hold the sediments, mixed with water, which are poured into the reinforced perimeter (see Figure 7.3b). The amount of sediment and water spread into the perimeter is calculated by considering the subsequent consolidation processes; thus, soon after spreading the artificial salt marsh has a quite high average elevation (+80/+100 cm above m.s.l.), and the average height gradually drops year after year. In the meantime, despite first being without vegetation, the first halophyte pioneer species colonize the terrain, and after 3/6 years, the vegetation cover is further diversified and the salt marsh starts to acquire greater complexity and year-by-year settles to the average characteristic height on the medium sea level (+30/40 cm above m.s.l.). Thanks to continuous interaction with high tides (submersion periods) and low tides (runoff processes), also the surface increases its diversification, with the formation of narrow tidal creeks and small pools of rain or seawater. The increased diversification attracts a higher variety of vegetation and several bird species, thus slowly restoring wetland habitats and ecosystems. From on-site monitoring, it has been observed that more than 10 years is needed to provide characteristics similar to those

Figure 7.3. (a) Example of salt marsh border protection against wave erosion made with natural materials (from Life VIMINE project website, https://www.lifevimine .eu/lifevimine.eu/index.html). (b) Salt marsh restoration (from Consorzio Venezia Nuova archive).

of a natural salt marsh. Finally, also the reinforced edge can be removed (Ministero Infrastrutture e Trasporti 2017).

7.3 MARSHLAND INVESTIGATION AND MONITORING

In the early 2000s, an investigation to characterize the hydro-geotechnical behavior of the shallowest sediments of the marshes subject to tide fluctuation was undertaken in order to deepen the knowledge of the consolidation process of the natural salt marshes and to estimate the compressibility indexes.

Laboratory and field investigations have concentrated on a typical lagoon marsh, located along the San Felice channel as shown in Figure 7.4, facing the fishing village of Treporti, where the field trial embankment described in Chapter 4 was constructed. The investigation consisted of soil profile reconstruction, permeability, and compressibility measurements, as well as monitoring of pore pressure evolution as a function of tide oscillation in the shallow subsoil of the marsh for a relatively long period (Cola et al. 2008).

In this lagoon area, the marshes are heavily subjected to the most typically significant environmental actions such as tidal flow, wave impact, wind, sedimentation, erosion, evaporation, etc., which cause a rapid evolution in marsh morphology. Figure 7.5 shows the soil profile down to around

Figure 7.4. Location of the test site and view of the marsh San Felice.

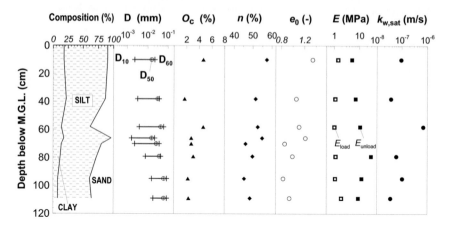

Figure 7.5. Soil properties down to 1 m depth (Cola et al. 2008).

1 m below marsh level. Geotechnical tests were carried out in order to determine the grain size distribution, natural water content w_0, organic content O_c, porosity, and in situ void ratio e_0. The soil is a mixture of sand, silt, and clay with a predominant silty fraction and a small content of clay. The mean diameter D_{50} is around 0.03 mm, while 10% of the material measures less than 0.003 mm in diameter. The organic material is composed of residuals of small root and plant fragments. The void ratio is relatively high compared to the typical values of the deeper soils and is conditioned by the presence of organic content. The last two columns summarize some experimental values of the stiffness modulus E and saturated hydraulic conductivity $k_{w,sat}$. The values of stiffness modulus E were estimated from oedometric tests in both loading (E_{load}) and unloading (E_{unload}) conditions.

Figure 7.6 shows a layout of the pore pressure measurement system installed in the San Felice marsh. The groundwater pressure was monitored

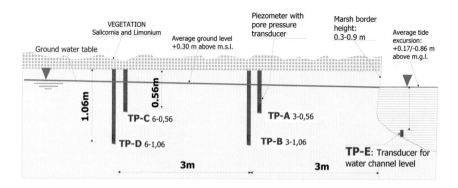

Figure 7.6. Groundwater pressure measurement system (Cola et al. 2008).

at two levels, that is at 0.56 m and 1.06 m, along two verticals spaced 3 m from each other and 3 m from the marsh border. A water pressure transducer was also installed in the nearby canal.

Groundwater pressure evolution was measured during spring. All the measurements were then referred to the mean marsh ground level (m.g.l.) assumed at +0.30 m above the mean sea water level (m.s.w.l.). A typical example of these measurements is provided in Figure 7.7a and b, which

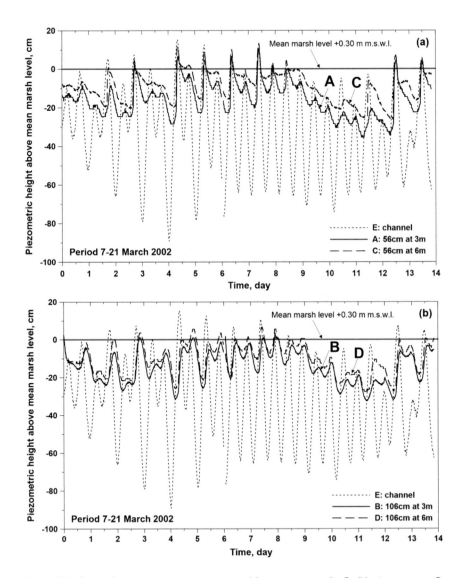

Figure 7.7. Groundwater pressure monitoring (a) piezometers A, C; (b) piezometers B, D (tide plotted with continuous line).

show the tide excursion and the pore pressure evolution measured in the ground over 14 days – from 7 to 21 March 2002 – during which the tide was characterized by cycles of submersion and emersion of the marsh surface. During that period evapotranspiration and rain had no impact on the pore pressure response because it never rained, air temperature was relatively constant, and vegetation was inactive.

It is interesting to note that the limited oscillation of pore pressure in the soil, during non-submerging tides, was characterized by significant damping and delay. The seepage gradient coupled with the low permeability shows that no significant groundwater drainage occurred in the marshy ground.

When the tide submerges the marsh, the pore pressure in the upper piezometers suddenly rises, reaching the same values of the tide, but subsequently the pore water trend is characterized by significant delay and damping. This phenomenon is due to the effect of partial saturation of the upper part of the marsh, as will be discussed in Chapter 9, which concerns the protection of the island of San Marco.

The observed phenomenon has been modeled using the finite element method to investigate the effect of cyclic pore pressure evolution on the stability of marsh scarps, as well. Details can be found in Cola et al. (2008).

As far as stability is concerned, the study revealed that in order to guarantee the stability of the scarp approaching verticality, a cohesive strength component is, of course, required. It should be kept in mind that in Venice lagoon silts, whose major contribution to shear strength is due to friction, a weak cohesion may be ascribed to several coexisting factors, such as capillary pressure in the aerated zone, plant roots, and extremely small cementation or diagenetic effects, which can easily vanish. It is shown (Cola et al. 2008) that accumulation of shear and volumetric deformation induced by the cyclic oscillation of shear and mean effective stresses may contribute to the decrease of cohesion, leading to a progressive failure that could propagate from the bottom of the scarp up to the tension crack observed in the shallower part of the marsh.

7.4 CHARACTERIZATION OF MARSHLAND COMPRESSIBILITY

Long-term survivability of coastal marshes is largely dependent on their capability to gain elevation at a rate comparable to the relative rise of sea level, thus maintaining their position within the intertidal zone (Kirwan et al. 2010). There is wide agreement that vertical accretion of coastal marshes is controlled by physical (e.g., sediment deposition during high tides), biological (e.g., plant productivity), and chemical (e.g., decomposition) processes (Morris et al. 2002; Kirwan and Megonigal 2013). Since these processes depend on the relative elevation between the marsh plat-form and the mean sea level, they are inherently linked to the rate of relative sea level rise (Fagherazzi et al. 2020). Recent coupled biomorpho-geome-chanical modeling of long-term marsh evolution highlights how relative sea level rise is the main factor forcing a coastal marsh to thicken (Zoccarato et al. 2019).

Natural compaction (also referred to as 'autocompaction') of shallow, recently deposited sediments forming the marsh subsoil is one compo-nent that contributes to the total relative sea level rise. Autocompaction increases the vulnerability of these ecosystems, outpacing gains from accre-tion. However, after the pioneering works by Cahoon and Reed (1995) and Cahoon et al. (2002) aimed at unraveling compaction and sedimentation from accretion records by means of surface elevation table devices, only recently some studies have demonstrated that subsurface processes exert significant influence on platform elevation in many wetland systems.

Monitoring records of coastal marsh accretion and autocompaction over yearly to decadal periods have been carried out in the Venice lagoon (Tosi et al. 2016; Da Lio et al. 2018; Zoccarato and Da Lio 2021) by using inte-grated instrumentation such as surface elevation table devices, multiple-depth leveling benchmarks, and persistent scatterer interferometry.

Thus, the evaluation of the marshland compressibility is fundamental to better model their long-term evolution because if autocompaction is improperly accounted for or neglected, inaccurate reconstructions of rela-tive sea level rise, land subsidence, and sedimentation rates are obtained (Morton et al. 2000; Serandrei-Barbero et al. 2006).

7.4.1 Experimental site testing

In order to better understand the consolidation processes and the compress-ibility characteristics of the natural salt marshes, experimental tests have been carried out in two different salt marshes of the Venice lagoon (Teatini et al. 2020; Zoccarato et al. 2022).

The loading experiments were carried out at Lazzaretto Nuovo (LN) and La Grisa (LG) marshes in the northern and southern basins of the Venice lagoon, respectively, as described in Figure 7.8. The elevation of the two marshes at the experimental site are 0.55 (LN) and 0.5 m (LG) above m.s.l.

To measure stress-strain-time shallow soil behavior the marsh surface was loaded with eight 500-l polyethylene tanks arranged in two rows of four tanks each, filled during loading cycles with seawater pumped from the nearest lagoon creek. The tanks rest on a reinforced geotextile and four wooden pallets to guarantee a uniform load distribution on the marsh sur-face and eliminate/reduce buoyancy forces on the tanks in high tide condi-tions. The maximum cumulative load, which is reached when all the tanks are filled, amounts to 40 kN and is distributed on a 4.0 m² surface. The

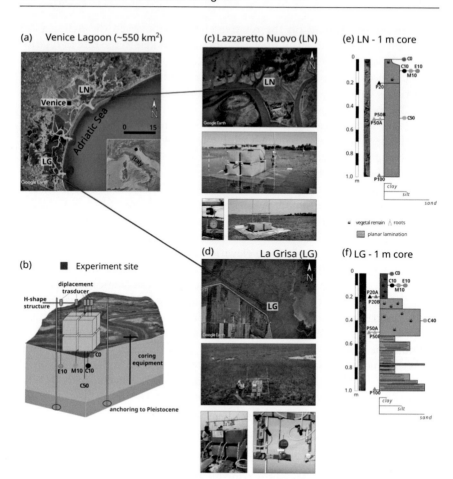

Figure 7.8 Map of the Venice lagoon with the locations of the LN and LG marshes where the loading experiment was carried out in July 2019 and October 2020, respectively. (b) Sketch of the experimental set-up showing the position of the monitoring instruments and the H-shaped structure used as a stable reference for the displacement transducers and anchored on the Pleistocene sediments. (c) and (d) Satellite images of the LN and LG marshes and photos taken during the ongoing experiments. At LN (c) the photos show the experimental area loaded with four tanks and eight tanks, in this latter case with the marsh platform submerged during high tide, and a detail of the instruments used to monitor vertical displacements. At LG (d) the photos show the experimental area with the eight tanks (drone photo by Rodrigo Gomila), a detail of the tanks during an emptying (unloading) phase, and the displacement transducers. (e) and (f) Stratigraphic profiles of the upper 1 m subsoil at LN and LG, respectively. Manual augering equipment was used to map the subsurface stratigraphy and collect undisturbed soil samples and cores of the marsh at the two sites in the close vicinity of the loading experiments to carry out a geotechnical laboratory characterization (Zoccarato et al. 2022).

designed configuration allows the transfer of the load to a sufficiently large area, thus the assumption of a nearly vertical one-dimensional strain condition is acceptable, at least below the central portion of the system. The response accounts for and averages local-scale heterogeneities, for example, those due to halophytic vegetation, of the marsh deposits, which cannot be avoided in few-cm size samples tested in the lab. Moreover, in an oedometric one-dimensional cell, the friction between the oedometer piston and cell does not permit low stress levels (i.e., those corresponding to a few cm on new sediments on a marsh surface), which are typically experienced by shallow deposits.

Instead, the field experiment, where the water-filled tanks provide the load, allows a compression test to be carried out at low vertical effective stresses, from 1–2 kPa up to a maximum of 10 kPa, that is stresses comparable to the low shallow overburden stress as well as that induced by the typical tide oscillation (5–6 kPa). Two to four loading and unloading phases, with an increasing percentage of tank filling, are carried out to characterize the marsh soil response in unloading and loading conditions and to separate recoverable from unrecoverable strain. To tentatively characterize the viscous response, the maximum load was maintained for a sufficiently long time, i.e., more than 24 hours.

The experiment is equipped with an appropriate monitoring system to measure vertical displacements and groundwater pore pressure at various depths and locations below the loaded area. As shown in Figure 7.9, the sensors are placed within the 0.2 m space left from the tank columns. In the planned configuration, five sensors measuring vertical displacements were used. Three of them were located in correspondence to the load center at the marsh surface (C0) and at 0.1 (C10) and 0.5 m (C50) below the marsh surface, the other two at the edge of the loaded area (E10) and in an intermediate position (M10) at 0.1 m depth. The sensor locations are designed to capture the shape of the soil deformation occurring below the tanks with maximum values occurring at the center of the load and decreasing as the edge is approached. Moreover, the maximum depth of the sensor (0.5 m below the surface) was selected to measure the main part of the expected deformation. A local benchmark network was also installed for an independent check of surface movements during the ongoing experiment. Four water pressure loggers are installed in the soil orthogonal to the displacement transducers (see Figure 7.8b) to measure the evolution vs. time of the water pressure induced either by the surface load or by the tide. The piezometer depths are 0.2 (P20A and P20B), 0.5 (P50A and P50B), and 1.0 m (P100). Similar to displacements, groundwater pressure is measured in correspondence with the load center, at the side of the loaded area, and in an intermediate position. Note in Figure 7.8 that an additional water pressure logger must be placed a few meters far from the experimental apparatus to record the tide oscillation.

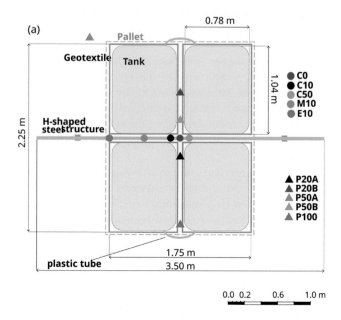

Figure 7.9 Plan of the loading experiment with dimensions, equipment (tanks, pallet, reference steel structure), location of the sensors to measure vertical displacements (bullets), and pore water pressure (triangles). The sensor coding is representative of the deployment depth (in cm) (Zoccarato et al. 2022).

7.4.2 Results

Figure 7.10 shows the results of the site test carried out at La Grisa, in terms of a dataset of vertical displacement and pore water pressure in marsh landforms. The displacements and pressures depend on the depth and location of the sensors and the loading or unloading phase.

The experiment lasted without interruption from 27 October to 2 November. The first loading phase reached 5.6 kPa (see Figure 7.10a), namely the four bottom tanks were filled, and the load was kept constant over 24 hours. The maximum settlement of the marsh surface (sensor C0) reached 10.2 mm (Figure 7.10b). C10 and M10 measure 4.4 and 3.4 mm, respectively (approximately 40% and 30% of C0). Settlement at C40 and E10 amounts to 1.1 and 0.5 mm, respectively (approximately 10% and 5% of C0). The displacement transducers showed a clear rebound following the tank emptying operated on 28 October at 4:00 PM. The load remained null for about 24 hours and was then increased to about 11.3 kPa and maintained for about 72 hours. A further settlement was recorded by all the displacement transducers (Figure 7.10b). C0 collected a final settlement of about 32 mm relative to the onset of the experiment. C10 and M10 measured a settlement up to 18 mm and 15 mm. The settlements were much

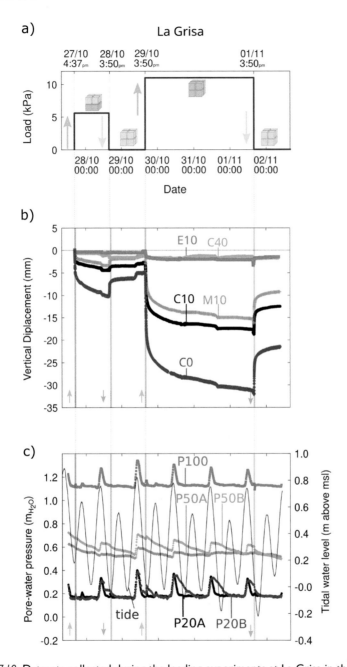

Figure 7.10 Datasets collected during the loading experiments at La Grisa in the Venice lagoon. (b) The load applied on the marsh surface following the various loading and unloading phases with the filling and emptying of the tanks. The filling and emptying last approximately 30–45 minutes. The loading experiments

(Continued)

were carried out from 27 October to 2 November in 2020 at LG. (d) Vertical displacement registered by each sensor versus time measured during loading and unloading phases. Negative values mean settlement; positive values mean uplift. C0, C10, and C50 (C40 in LG) are located directly below the load center and refer to the marsh surface, and 0.1 and 0.5 m (0.5 m) depth. E10 and M10 refer to a 0.1 m depth at the edge of the loaded area and in an intermediate position, respectively. (f) Pore water pressure (left axis) and tidal water level (right axis) were measured during the loading and unloading phases at various depths. P20, P50, and P100 refer to deployment depths equal to 0.2, 0.5, and 1.0 m below the marsh surface. Letters A and B identify sensors located on the two sides of the experiment, as indicated in Figure 7.9 (Zoccarato et al. 2022).

smaller at C40 and E10, amounting to 1 and 2 mm, respectively. Finally, all eight tanks were emptied, and the consequent rebound was recorded for about 24 hours. At the end of the experiment, the sensors accumulated a permanent displacement of about 22, 13, 10, 2, and 2 mm for C0, C10, M10, C40, and E10, respectively, corresponding to a settlement equal to 60–70% for C0, C10, M10, and E10, and 80% for C40, of the maximum displacement (Zoccarato et al. 2022).

The automatic logging of marsh displacements revealed additional insight into the displacement behavior. Small perturbations were observed in correspondence with the tide peaks, even though the water level on the marsh platform did not reach the tanks, except for the tide that occurred on 29 October at 11:00 AM (0.82 m above m.s.l). Moreover, the curves showed clear creep deformation (i.e., secondary, viscous deformation) following the load application after the excess pore-water pressure dissipation. Notice how the deformation rate (i.e., the slope of the displacement vs time curve) decreased with monitoring depth.

The behavior of the pore water pressure evolution is depicted in Figure 7.10c together with the tidal water level fluctuation. The pore water pressure followed the tidal level when the marsh was submerged, and the perturbation caused by the (un)loading operations amounts to a few centimeters (up to 0.04 m) and dissipated in about two to three hours. Note that the dissipation of the pore water pressure after the tide peaks followed the tidal evolution in P20A and P100, but dissipation considerably lagged in P20B, P50A, and P50B and that the sensors deployed at the same depth (i.e., P20A and P20B, and P50A and P50B) are characterized by quite different pressure behavior. The highly heterogeneous lithology distribution typical of the lagoon marshes, along both the vertical and horizontal directions, is likely responsible for this particular behavior of pore water pressure evolution.

The research project is ongoing and the interpretation of the collected dataset, integrated with laboratory measurements, may give proper insight into marsh subsurface response. The project is finalized to set up and calibrate a numerical model to predict the long-term settlement evolution due to the autocompaction of these highly heterogeneous soils.

The outcome of the research carried out at the San Felice marsh has been particularly useful for obtaining insight into the pore water pressure evolution induced by the tide in the shallowest marshy subsoil of the Venice lagoon, especially concerning the permanence for longer periods of time of a relatively higher average pore water pressure in the soil with respect to the average free lagoon water level. This distinct behavior helped to understand, some years later, the groundwater response of the subsoil in Piazza San Marco, to design suitable protection interventions, as will be shown in the next chapter.

REFERENCES

Allen, J. R. 2000. Morphodynamics of Holocene salt marshes: A review sketch from the Atlantic and Southern North Sea coasts of Europe. *Quaternary Science Reviews* 19(12), 1155–1231.

Amos, C. L., Umgiesser, G., Tosi, L. and Townend, I. H. 2010. The coastal morphodynamics of Venice lagoon and its inlets. *Continental Shelf Research* 30(8), 837–1018. https://doi.org/10.1016/j.csr.2010.01.014.

Cahoon, D. R., Lynch, J. C., Perez, B. C., Segura, B., Holland, R. D., Stelly, C., and Hensel, P. 2002. High-precision measurements of wetland sediment elevation: II. The rod surface elevation table. *Journal of Sedimentary Research* 72(5), 734–739.

Cahoon, D. R., Reed, D. J. and Day Jr., J. W. 1995. Estimating shallow subsidence in microtidal salt marshes of the southeastern United States: Kaye and Barghoorn revisited. *Marine Geology* 128(1–2), 1–9. https://doi.org/10.1016/0025-3227(95)00087-F.

Carniello, L., Defina, A. and D'Alpaos, L. 2009. Morphological evolution of the Venice lagoon: Evidence from the past and trend for the future. *Journal of Geophysical Research: Earth Surface* 114F04002, doi:10.1029/2008JF001157

Cola, S., Sanavia, L., Simonini, P. and Schrefler, B. A. 2008. Coupled thermo-hydromechanical analysis of Venice lagoon salt marshes. *Water Resources Research* 44, W00C05, doi:10.1029/2007WR006570.

Da Lio, C., Teatini, P., Strozzi, T. and Tosi, L. 2018. Understanding land subsidence in salt marshes of the Venice Lagoon from SAR interferometry and ground-based investigations. *Remote Sensing of Environment* 205, 56–70. https://doi.org/10.1016/j.rse.2017.11.016.

D'Alpaos, A., Lanzoni, S., Mudd, S. M. and Fangherazzi, S. 2006. Modeling the influence of hydroperiod and vegetation on the cross-sectional formation of tidal channels. *Estuarine, Coastal and Shelf Science* 69, 311–324.

Fagherazzi, S., Mariotti, G., Leonardi, N., Canestrelli, A., Nardin, W. and Kearney, W. S. 2020. Salt marsh dynamics in a period of accelerated sea level rise. *Journal of Geophysical Research: Earth Surface* 125(8), e2019JF005200.

Fagherazzi, S., Palermo, C., Rulli, M. C., Carniello, L. and Defina, A. 2007. Wind waves in shallow microtidal basins and the dynamic equilibrium of tidal flats. *Journal Geophysical Research* 112, F02024.

Fletcher, C., Da Mosto, J. and Rotondo, M. 2004. In U. Allemandi (ed.), *La scienza per Venezia: Recupero e salvaguardia della città e della laguna*. Torino, Italy (*in Italian*).

Guerzoni, S. and Tagliapietra, D. 2006. *Atlante della Laguna Venezia tra Terra e Mare*. Venezia: Marsiglio Editori. (*in Italian*).

Helsby, R., Amos, C. L. and Umgiesser, G. 2006. *Morphological Evolution and Sand Pathways in Northern Venice Lagoon, Italy*. Scientific Research and Safeguarding of Venice, Research Programme 2004–2006, vol.5. CORILA, Venice (388-329.402).

Kirwan, M. and Megonigal, P. 2013. Tidal wetland stability in the face of human impacts and sea-level rise. *Nature* 504, 53–60. https://doi.org/10.1029/2011RG000359.

Kirwan, M. L., Guntenspergen, G. R., d'Alpaos, A., Morris, J. T., Mudd, S. M. and Temmerman, S. 2010. Limits on the adaptability of coastal marshes to rising sea level. *Geophysical Research Letters* 37(23). https://doi.org/10.1029/2010GL045489.

Life VIMINE (Venice Integrated Management of Intertidal Environment) Project, Co-Funded by the European Commission through the LIFE+ Nature Programme (2013–2017). *An Integrated Approach to the Sustainable Conservation of Intertidal Salt Marshes in the Lagoon of Venice* (Grant Agreement Life 12 NAT/IT/001122). https://www.lifevimine.eu/en/index.php.

Ministero delle Infrastrutture e dei Trasporti – Provveditorato Interregionale per le Opere Pubbliche del Triveneto – Consorzio Venezia Nuova (2017) *Tecnologia, sviluppo e innovazione per la difesa ambientale e costiera*.

Morris, J. T., Sundareshwar, P. V., Nietch, C. T., Kjerfve, B. and Cahoon, D. R. 2002. Response of coastal wetlands to rising sea level. *Ecology* 83(10), 2869–2877. https://doi.org/10.1890/0012-9658(2002)083[2869:ROCWTR]2.0.CO;2.

Morton, R. A., Ward, G. H. and White, W. A. 2000. Rates of sediment supply and sea-level rise in a large coastal lagoon. *Marine Geology* 167(3–4), 261–284. https://doi.org/10.1016/S0025-3227(00)00030-X.

Perillo, G., Wolanski, E., Cahoon, D. R. and Hopkinson, C. S. (Eds.). 2018. *Coastal Wetlands: An Integrated Ecosystem Approach*. Amsterdam, The Netherlands: Elsevier.

Serandrei-Barbero, R., Albani, A., Donnici, S. and Rizzetto, F. 2006. Past and recent sedimentation rates in the Lagoon of Venice (Northern Italy). *Estuarine, Coastal and Shelf Science* 69(1–2), 255–269. https://doi.org/10.1016/j.ecss.2006.04.018.

Tambroni, N., Stansby, P. K. and Seminara, G. 2005. Modelling the morphodynamics of tidal inlets. In C. A. Fletcher and T. Spencer (Eds.), *Flooding and Environmental Challenges for Venice and its Lagoon: State of Knowledge*. Cambridge: Cambridge University Press, 379–389.

Teatini, P., Da Lio, C., Tosi, L., Bergamasco, A., Pasqual, S., Simonini, P., Zambon, G. 2020. Characterizing marshland compressibility by an in-situ loading test: Design and set-up of an experiment in the Venice Lagoon. *Proceedings of the International Association of Hydrological Sciences* 382, 345–351.

Tommasini, L., Carniello, L., Ghinassi, M., Roner, M. and D'Alpaos, A. 2019. Changes in the wind-wave field and related salt-marsh lateral erosion: Inferences from the evolution of the Venice Lagoon in the last four centuries. *Earth Surface Processes and Landforms* 44(8): 1633–1646.

Tosi, L., Teatini, P., Carbognin, L. and Brancolini, G. 2009. Using high resolution data to reveal depth-dependent mechanisms that drive land subsidence: The Venice coast, Italy. *Tectonophysics* 1(474), 271–284.

Tosi, L., Da Lio, C., Strozzi, T. and Teatini, P. 2016. Combining L- and X-Band SAR interferometry to assess ground displacements in heterogeneous coastal environments: The Po River Delta and Venice Lagoon, Italy. *Remote Sensing* 8, 308.

Umgiesser, G., De Pascalis, F., Ferrarin, C. and Amos, C. L. 2006. A model of sand transport in Treporti channel: Northern Venice lagoon. *Ocean Dynamics* 56, 339–351.

Zoccarato, C. and Da Lio, C. 2021. The Holocene influence on the future evolution of the Venice Lagoon tidal marshes. *Communications Earth & Environment* 2(1), 1–9.

Zoccarato, C., Da Lio, C., Tosi, L. and Teatini, P. 2019. A coupled biomorpho-geomechanical model of tidal marsh evolution. *Water Resources Research* 55(11), 8330–8349. https://doi.org/10.1029/2019WR024875.

Zoccarato, C., Minderhoud, P. S. J., Zorzan, P., Tosi, L., Bergamasco, A., Girardi, V., Simonini, P., Cavallina, C., Cosma, M., Da Lio, C., Donnici, S. and Teatini, P. 2022. In situ loading experiments reveal how the subsurface affects coastal marsh survival. *Communications Earth & Environment* 3, 264. https://doi.org/10.1038/s43247-022-00600-9.

Chapter 8

Long-term behavior of historical building foundations

8.1 INTRODUCTION

The historic center of Venice is made up of several closely spaced islands located in the middle of the lagoon. The islands were initially reinforced and artificially stabilized to support the Venetian citizens' houses, buildings, storehouses, and naval yards. Many historically significant buildings were founded on flat wooden 'rafts', laid at a depth of 2–3 m (referred to as *zatteroni*). In the case of substantial loads, the rafts were placed on top of small in both diameter and length wooden piles that were embedded at a shallow depth. Wooden piles were frequently used for external masonry walls and heavily stressed foundations, such as those supporting quay walls, bridges, bell towers (e.g., Gottardi et al. 2015), or tall buildings, while *zatteroni* were prevalently utilized for interior walls.

Figure 8.1 is a charming watercolor painted in the 18th century by Giovanni Grevembroch (1731–1807), a Venetian painter of Flemish origin, and shows how wooden piles were driven into the ground. The photograph in Figure 8.2 clearly illustrates the purpose of wooden piles used for land reclamation.

In fact, small wooden piles improve the mechanical behavior of the upper soft silty clay, which characterizes the shallowest layer of the Venice lagoon, as already described in Chapters 2 and 4. Since the piles are rather short and tightly spaced, the bearing capacity provided by the improved soil only slightly increases but does reduce expected settlements considerably.

One of the most famous examples of the broad use of wooden piles as supporting elements is offered by the stone arch Rialto Bridge, which crosses the Grand Canal in the middle of the historic city. This method of foundation soil improvement, however, has been used extensively by Venetian architects and engineers (e.g., Lazzarini 2006).

Giovanni Stringa provided a contemporaneous description of the construction of the new Rialto Bridge, completed in 1604 (reported in Lazzarini 2018), which substituted the previous one:

> *Onde dato principio a disfar il vecchio l'anno 1587 a di primo febraro,*
> *si comincio pur in detto anno et giorno anco a cavar il terreno per le*

DOI: 10.1201/9781003195313-8

Figure 8.1 Watercolor of wooden pile driving in Venice (watercolor by Grevembroch 1780).

fondamente del nuovo, et le cavarono per piedi sedici sotto, et quivi poscia piantarono dodici mila palli di olmo, lunghi piedi dieci l'uno, sei mila overo in circa per ciascuna parte, cosi di qua come di la del canale; et erano cosi spessi che si toccavano insieme: vi posero poscia sopra quelli un suolo di tavoloni di larese a traverso un sopra l'altro, di grossezza poco meno d'un palmo; poi vi accommodarono per ogni canto bordonali pur di larese, lunghi piu di quaranta piedi l'uno: dopo tutte queste cose, viste da me coi proprij occhi, fu messa la prima pietra a di 9 giugno 1588.

A translation from the Venetian local language may be as follows [translation by author]:

Once the destroying of the old bridge had started on the first of February 1587, in that same year the soil was removed down to sixteen

Figure 8.2 View of wooden piles as settlement reducing piles (photo by G. Arici).

feet (1 feet = 0.3477 m) to set up the foundations of the new bridge, and then twelve thousand elm piles were driven, each with a length of ten feet, six thousand for both sides of the canal; they were so dense that they touched each other. Then large crossed larch planks were laid above, slightly less than one palm (1 palm = 0.2486 m) thick; then larch beams, each more than forty feet long, were installed. After all these things I have seen with my own eyes, the first stone was placed on the first of June 1588.

The above statement represents a very concise and effective description of a moment that remained for centuries in the collective memory and can be found in many books and publications as well as on several websites.

Figure 8.3 is a drawing of the Rialto Bridge construction scheme attributed to the architect Luca Beltrami. Beltrami's design is characterized by a unique arch with a span of 28 m; note the 12,000 closely spaced elm piles that sustain the two shoulders of the bridge.

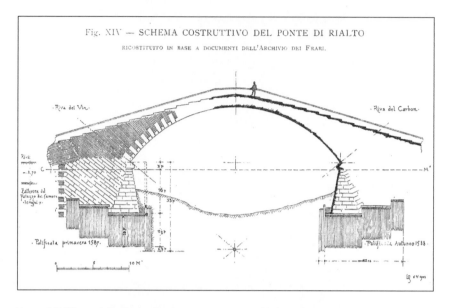

Figure 8.3 View of the Rialto Bridge construction scheme referred to as *Disegno Beltrami* (from Lazzarini 2018).

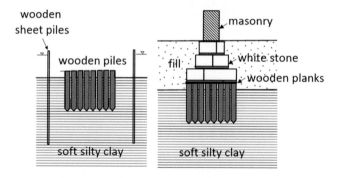

Figure 8.4 Construction sequence of an ancient wooden pile foundation in Venice: sheet piling, dewatering, and wooden pile driving (left); plank installation, calcareous stones, soil filling material, and masonry wall (right).

The construction sequence used by Venetian citizens to improve the soil may be basically described as follows (see also Figure 8.4):

- Wooden sheet pilings were initially driven into the surrounding ground to prepare a dry working area (in the local language referred to as *tura* from the verb *turare* = to plug), and then improving wooden piles were installed by percussion as shown in Figure 8.1;

- A horizontal surface was then prepared by evening off the heads of the piles and laying down wooden planking on which white calcareous stone blocks that supported the above masonry walls were placed;
- While the construction was proceeding, remolded soil was filled back into the empty area, up to the ground floor level. During this construction stage, some consolidation took place resulting in a slight improvement in the undrained soil shear strength, before the construction of the rest of the building (Ceccato et al. 2014; Bettiol et al. 2016).

8.2 WOODEN PILE DEGRADATION IN ANOXIC CONDITION

The structure of wood is made up of cells, whose walls are composed of micro-fibrils of cellulose and hemicellulose, embedded in a matrix of lignin. This particular structure results in a highly heterogeneous and anisotropic material.

Animals as well as fungi, bacteria, and chemicals can modify the original material's properties and cause the wood to decay. Wooden piles used as foundation elements lying above the groundwater table are subject to attacks by fungi and, in some particular conditions, may lose their bearing function in a handful of years. When the wood is partially submerged in salt water, it tends to be attacked by marine organisms, such as shipworms, which especially act in the tidal excursion zone. It is a common assumption that wooden piles totally submerged in water, that is, in absence of possible oxidation processes, can last indefinitely.

On the contrary, recent research has shown that some bacteria can deteriorate wood over a very long term, even in anoxic conditions, by attacking the cell wall. The wood preserves its original volume when wet but can suddenly collapse when extracted from the water and exposed to air for drying (Peek and Willeitner 1981). This result has been confirmed by chemical and mechanical analyses of wood samples taken from Venetian foundations, which were, in some cases, in an advanced degradation state (Biscontin et al. 2009; Gottardi et al. 2015).

Several types of bacteria are always present in wood surrounded by soil, but its degradation rate depends on the type of wood and the environmental conditions (Kretschmar et al. 2008; Klaassen 2008). The outer part of a wooden pile is usually characterized by the presence of juvenile wood, namely the sapwood layer, more prone to bacteria attack. Klaassen and Van Overeem (2012) observed that in many pine piles used as foundation elements in the Netherlands, the whole sapwood layer was moderately to severely degraded. The level of degradation decreases moving both inward and downward, as the outer part and head of the pile are generally in worse conditions than the rest (i.e., the hardwood).

Klaassen and Van Overeem (2012) reported, in some cases, penetration rates of the degradation front into the pile of 0.13 mm/year for spruce and 0.25 mm/year for pine, but also much less or negligible deterioration rates have been recorded. Extending these observations to piles with a diameter of 150–200 mm, the entire pile section may decay in about seven to eight centuries.

The evaluation of the state of wood conservation may be based on some chemical-physical parameters, the most important being:

- Basal density (BD), which is the ratio between dry mass and wet volume, sometimes referred to as the typical value of fresh wood, obtaining residual basal density (RBD = $BD_{deteriorated}/BD_{fresh\,wood}$);
- Maximum water content (MWC), which is the ratio between the mass of water and the anhydrous mass of wood;
- Cellulose content (H) and lignin content (L).

These parameters are strongly correlated in deteriorated wood, all of them causing a reduction in wood strength and stiffness.

The BD decreases over time as a function of MWC increases (according to a power law that is independent of the species), which, in turn, causes a reduction in wood strength and stiffness. The BD decreases exponentially as the MWC increases, depending on the relationship between these variables in the type of wood.

Figure 8.5 shows some results obtained from laboratory testing carried out on wood samples taken from the foundations of a late medieval quay wall, facing the San Felice Canal. The wall was subjected to dewatering, necessary for wall retrofitting. The wood belongs to some species typically used in Venetian foundations such as elm, larch, spruce, oak, and pine. The trend is characterized by a nonlinear relationship between MWC and BD, that is, as MWC increases, BD decreases. It is important to note that the MWC differs greatly among the tested samples, indicating that the same type of wood can decay at different times, depending on its position.

It is also observed that wood volume V, at changing MWC and BD, remains approximately constant under groundwater level (V = V_0, V_0 = initial wood volume), even during its degradation process. This leaves some layers of the cell wall intact, these being strong enough to keep waterlogged wood in its original shape (Huisman et al. 2007).

Bacteria predominantly decompose the cellulose that is replaced by water, while the percentage of lignin remains constant. This means that when BD and H/L decrease while MWC increases, the wood's mechanical properties are subject to a significant reduction. A trend of possible evolution of these parameters is shown in Figure 8.6.

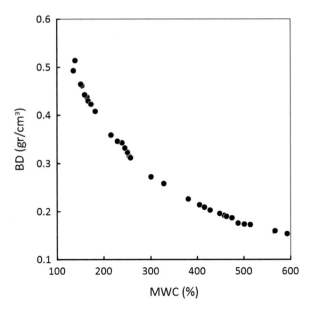

Figure 8.5 Relationship between MWC and BD for ancient Venetian wood.

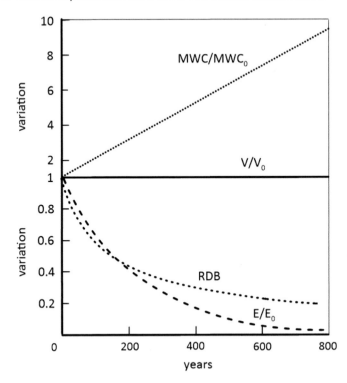

Figure 8.6 The trend of decay of wood properties over centuries (Ceccato et al. 2014).

8.3 PILE DEGRADATION MODELING

To characterize the effects of wood degradation on the overall behavior of the wooden pile–soft soil system, a simple constitutive model for wood is used. The model does not account for all the factors influencing the wood's mechanical behavior but is basically limited to the description of the wood degradation process. The selected constitutive equations are those of the linear isotropic elasticity – perfect plasticity with Tresca yield criterion, namely it is assumed for simplicity that wood behaves like a purely cohesive material, with strength and stiffness changing over time according to degradation processes (Ceccato et al. 2014).

Due to the lack of studies providing relationships describing the decay of mechanical parameters over time, a simple linear decrease in strength and stiffness is used.

Since measurements of stiffness in ancient wood are not available, it is assumed that strength and stiffness, frequently proportional in wood, may decrease according to the same law:

$$\zeta = \frac{E}{E_0} = \frac{\sigma_c}{\sigma_0} \tag{8.1}$$

where E_0 is the elastic modulus of fresh wood and σ_c and σ_0 are the compression strength of aged and fresh wood, respectively, while ζ represents the degradation level.

The single pile behavior is studied in detail to appreciate the variation of load distribution as a function of wood deterioration. To model the effects of the degradation of a single pile on the overall behavior, an axisymmetric FE model was set up, characterized by a pile of 3 m in length and a diameter of 20 cm as shown in Figure 8.7a and b.

According to the Mohr–Coulomb elasto-perfectly plastic material model, the soil properties are friction angle $\phi = 30°$; cohesion $c = 7$ kPa, Young's modulus $E = 30$ kPa, Poisson's ratio $\nu = 0.30$. Young's modulus of fresh pine wood is $E_{wood} = 7,000$ MPa. The pile–soil interface cohesion and the interface friction angle have been assumed as 70% of those of the soil.

After the K_0 geostatic stress initialization, the pile and overlying raft were activated by applying a load of 200 kPa. Wood degradation was simulated in steps by reducing stiffness and strength, the worst situation corresponding to a reduction of 1/1,000 of the original pile parameters.

Figure 8.8 shows the ratio between the force transferred to the pile (F_{pile}) with respect to that of the soil (F_{soil}). It is noteworthy that a significant change in the overall response is observed only with the ratio ζ well beyond 0.1. This is due mainly to a strong reduction in shaft resistance $F_{pile\,shaft}/F_{soil}$ as shown in Figure 8.8 brought about by the progressive wood degradation.

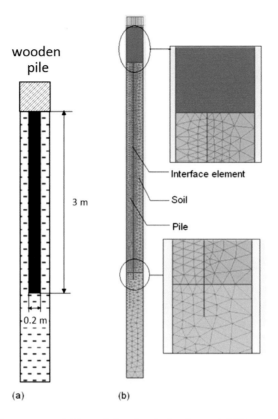

wooden
pile

3 m

0.2 m

(a)

(b)

Interface element

Soil

Pile

Figure 8.7 (a) Geometry and (b) mesh for the single pile model (Ceccato et al. 2014).

8.4 LONG-TERM BEHAVIOR OF WOODEN PILE GROUP

Load transfer from the foundation to the soil, enhanced by the wooden piles, occurs proportionally to the relative pile/soil stiffness. When wood deteriorates and loses stiffness, stress transfers to the soil and deformations arise in the entire reinforced zone.

The effects of wood degradation depend on the characteristics of the piles (type of wood, length, diameter) and soil profile as well as on the relative pile density α given by the ratio between the volume of soil and the volume of piles.

To account for the coupled effect of wood degradation and time-dependent soil behavior, a two-dimensional FE model has been used (Bettiol et al. 2016). A strip piling disposition is modeled in plane strain condition using Plaxis 2D (Brinkgreve 2011). As represented in Figure 8.9, the overall foundation is 2.40 m wide and is composed of 2 m long piles, floating in

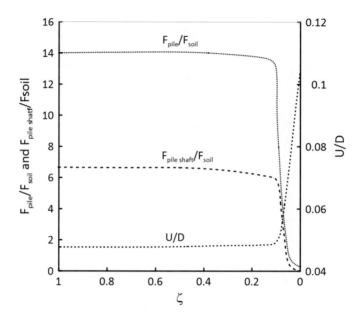

Figure 8.8 Load transfer and settlement as a function of wood decay (Ceccato et al. 2014).

5 m thick and soft cohesive soil, described using the Soft Soil Creep model (SSC) (Vermeer and Neher 1999), overlying a sandy layer down to the model boundary. SSC is used together with the constitutive model described previously for the properties of decaying wood; the mechanical properties for both types of Venetian soils have been estimated on the basis of the wide experimental research described in Chapter 4.

The single piles have been simulated with clusters and interfaces. Wood decay is described by a linear decrease in the elastic modulus (E) and compressive strength (σ_c) from the original values to zero in 300 years after construction (taken equal to 1 year); simulations were stopped when $E/E_0 = \sigma_c/\sigma_0 = 0.3\%$, a value that may highlight the main features of the load transfer phenomenon.

The elastic modulus $E_0 = 5{,}000$ MPa and the compressive strength $\sigma_0 = 12$ MPa correspond to the mean values for fresh pine, reduced by 50%, which is used to consider damage during driving and accounting for the defects of wood.

Calculation steps attempted to describe the main features of the transfer process of loading, considering both wood decay and soil creep.

Initially a uniformly distributed load equal to 35 kPa was applied to simulate the weight of the filler material layer; then the construction followed, turning a load of 100 kPa over the piles, and finally wood deterioration was represented by changing the piles' material properties.

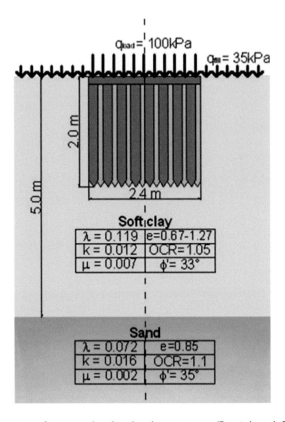

Figure 8.9 Geometry of improved soil and soil parameters (Bettiol et al. 2016).

Figure 8.10 depicts the time-displacement curves for some values of the coefficient of substitution α, where $\alpha = 100\%$ represents the theoretical limit case of a fully wooden foundation. High substitution ratios (60%–80%) are typical of Venetian wooden pile improvement systems.

From Figure 8.10, it may be noted that overall settlements are mainly due to the time-dependent behavior of soil, until the wood has almost entirely lost its mechanical properties.

The steepest part of the curves ($E/E_0 < 1.7\%$) means that the wood is in a fully plastic state, but the collapse is avoided thanks to the carrying capacity of the soil between piles. If lower values of wood strength are considered, then higher displacement will be seen.

From the calculation carried out so far, it appears that the most important process controlling foundation displacement is the creep behavior of the shallowest Venice soils. This was already observed in Chapter 3, when the natural soil compression due to natural subsidence was presented and discussed. Nevertheless, the degradation of wooden piles in anoxic condition

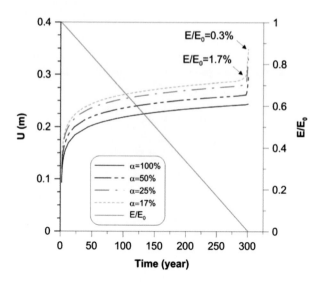

Figure 8.10 Time–displacement curves as a function of substitution ratio (Bettiol et al. 2016).

up to negligible stiffness and strength may contribute to a sudden increase in settlements, especially in the case of heavily stressed foundations, such as in the case discussed by Gottardi et al. (2015).

REFERENCES

Bettiol, G., Ceccato, F., Pigouni, A. E., Modena, C. and Simonini, P. 2016. Effect on the structure in elevation of wood deterioration on small-pile foundation: Numerical analyses. *International Journal of Architectural Heritage* 10(1), 44–54.

Biscontin, G., Izzo, F. and Rinaldi, E. 2009. *Il sistema delle fondazioni lignee a Venezia: Valutazione del comportamento chimico-fisico e microbiologico.* Venice, Italy: CORILA Publication. (*in Italian*).

Brinkgereve, R. B. J., Swolfs, W. M. and Engin, E. 2011. *PLAXIS 2D 2011–User Manual.* Delft, the Netherlands: Plaxis BV.

Ceccato, F., Koeppl, C., Simonini, P., Schweiger, H. F. and Tschuchnigg, F. 2014. FE analysis of degradation effect on the wooden foundations in Venice. *Rivista Italiana di Geotecnica* 2, 27–37.

Gottardi, G., Lionello, A., Marchi, M. and Rossi, P. P. 2015. Monitoring-driven design of multiphase intervention for the preservation of the Frari Bell Tower in Venice. *Rivista Italiana di Geotecnica* 1, 45–64.

Huisman, D. J., Manders, M., Kretschmar, E., Klaassen, R. K. W. M. and Lamersdorf, N. 2007. Burial conditions and wood degradation on archaeological sites in the Netherlands. *International Biodeterioration & Biodegradation* 61(1), 33–44.

Klaassen, R. K. W. M. 2008. Bacterial decay in wooden foundation piles— Patterns and causes: A study of historical pile foundations in the Netherlands. *International Biodeterioration & Biodegradation* 61(1), 45–60.

Klaassen, R. K. W. M. and Van Overeem, B. S. 2012. Factors that influence the speed of bacterial wood degradation. *Journal of Cultural Heritage* 13(3), S129–S134.

Kretschmar, E. I., Gelbrich, J., Militz, H. and Lamersdorf, N. 2008. Studying bacterial wood decay under low oxygen conditions – Results of microcosm experiments. *International Biodegradation & Biodeterioration* 61(1), 69–84.

Lazzarini, A. 2006. *Palificate di Fondazione a Venezia, La Chiesa della Salute.* Archivio Veneto, s. V, Vol. CLXXI (2008), 33–60. (*in Italian*).

Lazzarini, A. 2018. *Legno e pietra. Sottofondazioni e fondamenta del ponte di Rialto.* Archivio Veneto, Sesta Serie, N. 16, 2018. (*in Italian*).

Peek, R. D. and Willeitner, H. 1981. Behaviour of wooden pilings in long time service. *Proceedings 10th International Conference of Soil Mechanics and Foundation Engineering, Stockholm*, 3, 147–152.

Vermeer, P. A. and Neher, H. P. 1999. A soft soil model that accounts for creep. In *Proceedings of the International Symposium—Beyond 2000 in Computational Geotechnics: Ten Years of PLAXIS International* (March 18–20, 1999, Amsterdam, The Netherlands). Rotterdam, Amsterdam: Balkema, 249–261.

Chapter 9

Safeguarding Piazza San Marco

9.1 INTRODUCTION

Piazza San Marco is the main square of the historical city of Venice and several historical buildings of high artistic and historic value rise in its proximity (Figure 9.1). On the east side of the square stand Basilica of San Marco, the Ducal Palace, and the Patriarchal Palace. The campanile is situated in front of the basilica, to the southwest. The southeast side of the area faces the basin of San Marco, while the Biblioteca Marciana (San Marco's library) and the Procuratie Nuove (formerly where the Procurators of San Marco, high officers of state during the Republic of Venice, lived and worked) rise on the southwest side. The Napoleonic wing and Procuratie Nuove delimit the square on the west and northwest sides, respectively. Several shops, cafés, and restaurants are situated at the ground level of these buildings. The small square on the north side of the church is known as Piazzetta dei Leoncini and just beyond that is the Torre dell'Orologio, the clocktower facing Piazza San Marco.

This area is the lowest part of the city, with an average elevation between +0.80 m and +0.90 m above the sea level at Punta della Salute (s.l.P.S.), and it is frequently flooded during high tides, as already described in Chapter 1. Flooding induces considerable deterioration of masonry walls, foundations, and decorative architectural elements of buildings (Fletcher and Spencer 2005; Ceccato et al. 2014; Bettiol et al. 2016), jeopardizing the historical heritage. When the water level reaches +0.60 m, it inundates a small area in front of the Basilica and its narthex. With a water level of +0.90m, approximately 65% of the square is flooded, and with a level of +1.15 m, it is completely submerged. A water level higher than +0.90 m is currently observed for 0.9% of the time on an annual basis, but this percentage will increase in the future due to the effects of subsidence (−0.5 mm/year) and an increase in mean water level due to climate change, as discussed in Chapter 3.

As introduced in Chapter 6, the activation of the MoSE gates occurs only when a sea level exceeding +1.10 m s.l.P.S. or higher is forecast, thus flooding of the historical area of San Marco is still possible, and specific protection measures are necessary. Finding an optimal solution in this peculiar

DOI: 10.1201/9781003195313-9

Figure 9.1 View of San Marco's Island with the main historic buildings (modified from GoogleEarth, Google 2021).

context is challenging as it must be respectful of the historical heritage, compatible with the touristic and economical activities of the most important site of the city, as well as cost-effective.

The main flooding mechanism of the square and possible solutions are summarized in Table 9.1 and Figure 9.2. In this chapter, mechanism 4 is

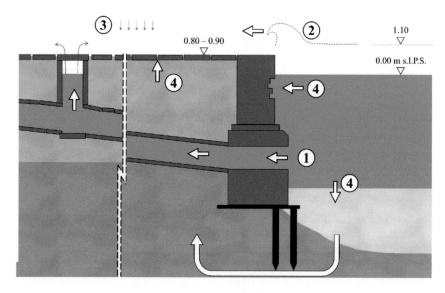

Figure 9.2. Main flooding mechanisms of San Marco's Square: (1) Back-flow through the drainage system, (2) overtopping, (3) heavy rainfall, (4) seepage through the soil.

Table 9.1 Mechanisms of flooding and possible interventions

	Mechanism	Possible solution
1	Back-flow through the drainage system, which is hydraulically connected with the lagoon	• Installation of special valves to close the entire drainage system in correspondence with very high tide • Restoration of the old *gatoli*
2	Overtopping and overflow from the quay walls	• Increase the elevation of the boundary walls of the area • Installation of special permanent floating barriers at some distance from the quay wall facing the San Marco's basin to control overtopping due to wind-induced waves
3	Heavy rainfalls occurring simultaneously with high sea levels	• Installation of an automatic pumping system removing the rainfall water collected by the drainage network into a new storage tank
4	Seepage through the soil below the paving and the permeable joints between the stone elements	• Impermeable membrane laid horizontally below the paving • Impermeable vertical sheet pile walls around the perimeter

discussed in detail, while mechanisms 1, 2, and 3 have been investigated in specific maritime and hydraulic studies (Salandin 2020; Ruol et al. 2020).

Water can enter the square by infiltrating through the open joints between the stones of the pavement, which are hydraulically connected to the basins through the subsoil. In order to avoid this seepage mechanism, vertical and horizontal impermeabilization measures are considered. The construction of vertical cut-off sheet pile walls around the boundary of San Marco's Island can prevent water from infiltrating through and below the quay walls. Moreover, an impermeable membrane inserted below the stone pavement can avoid seepage flow through the permeable joints between the stone elements of the pavement. These are, however, extremely costly interventions, which are recommended only if a significant amount of water is expected to enter by mechanism 4, which would be difficult to drain out with other techniques. The geotechnical study presented in the following provides insight into this issue.

9.2 THE DRAINAGE NETWORK

The drainage system of the historic city was mostly designed and constructed in the 18th century by the Republic of Venice and is still operating. It is composed of a network of masonry tunnels, locally named *gatoli*, that collect both rainfall and wastewater.

The typical sections of these *gatoli* are shown in Figure 9.3; they commonly feature a rectangular section with a width that can vary between

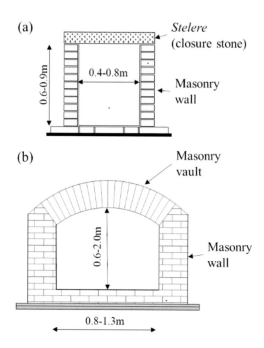

Figure 9.3 Typical cross section of Venetian *gatoli*.

0.4m and 0.8 m and a height between 0.6 and 0.9 m. Lateral walls are about 0.30 m thick, and a closure stone, referred to as stelere, ends the section. Wider sections feature a masonry vault and can reach a width of 1.50 m and a height of 2 m (see Figure 9.3b). Due to this construction system, the *gatoli* cannot be considered impermeable, since water can infiltrate through the fissures between the upper stelere and the lateral walls, and, in some cases, the mortar between masonry elements has significantly deteriorated. Figure 9.4a and b shows a view of the interior of one conduct in good condition and one in worse condition.

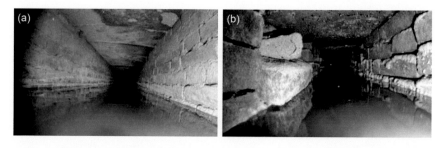

Figure 9.4 Inside view of two drainage conducts in San Marco's square: (a) *gatolo* in good condition, (b) *gatolo* in worse condition.

The conduits gently slope toward the canals; thus, the water velocity is generally quite slow. This creates sediment deposition, which can reduce their effective section by 50%–75% over time or, in extreme cases, completely obstruct the section (Volpato 2019). Hydraulic measurements and models (Volpato 2019) show that the hydraulic head inside the gatoli coincides with the sea water level, i.e., the system is well connected to the sea and water flows in and out according to tidal fluctuation.

Since the *gatoli* are not impermeable and are connected to the sea, they represent important elements of interaction between the groundwater and the lagoon.

9.3 SOIL PROFILE OF PIAZZA SAN MARCO

Three main geotechnical campaigns have been carried out in Piazza San Marco in order to define the geotechnical model of the subsoil. In 1993, the following geotechnical investigations were conducted: 6 geotechnical boreholes (16 m deep) with the collection of undisturbed samples, 10 standard penetration tests (SPT), 6 Lefranc permeability tests, and 3 piezocone tests (CPTU) with dissipation tests. In 1997–1998, 6 geotechnical boreholes up to 20 m or 32 m depth (with undisturbed sample collection), 9 SPT, 8 CPTU, and Boutwell in situ permeability tests were carried out. The results of these surveys highlighted the spatial heterogeneity of the subsoil; in particular, the hydraulic conductivity can vary by several orders of magnitude in the superficial anthropic layer.

Given the complexity of the system, this information was enhanced by new data collected in 2019. In particular, the most recent geotechnical site investigation consisted of geotechnical boreholes up to a depth of 20 m, with the collection of undisturbed samples; CPTU, SCPTU, DTM, and SDTM were also driven 20 m deep. Geotechnical laboratory tests included classification tests (grain size distribution, volume unit weight, water content, Atterberg limits, etc.), permeability tests in triaxial cells, with constant head and variable head permeameters, and one-dimensional oedometric compression tests. The locations of the tests are shown in Figure 9.5.

The position, dimensions, and state of preservation of the elements constituting the drainage system were assessed by correlating archaeological data with the results of electric tomography and Georadar investigations as well as performing video inspections.

The stratigraphy of Piazza San Marco is characterized by a superficial anthropic layer (formation 1/1A) with a thickness of 3.0–4.5 m, which underwent a number of anthropic actions over the centuries, reflected in its heterogeneity (Bortoletto 2019). Under the anthropic layer, a low permeability layer (from sandy silt (2) to clayey silt (2A)) has been identified, with a thickness between 2.0 m and 7.0 m. Beneath this, there is a more

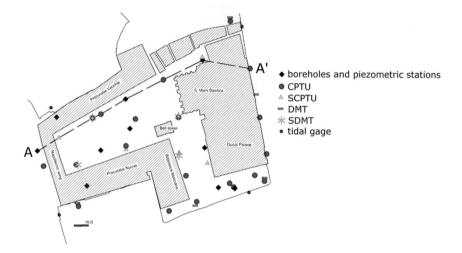

Figure 9.5 Position of in situ geotechnical tests (CPTU, SCPTU, DMT, and SDMT), bore-holes, and piezometric verticals.

permeable sandy formation (3 or 3A), whose thickness varies from 0.50 m to approximately 8.00 m. From the depth of 9.80 m below s.l.P.S., up to the maximum sounding depth, there is the typical alternation of prevalently clayey layers (4A) with lenses of moderately silty sand (4). A typical stratigraphy is shown in Figure 9.6 and a summary of the main features of soil layers can be found in Table 9.2.

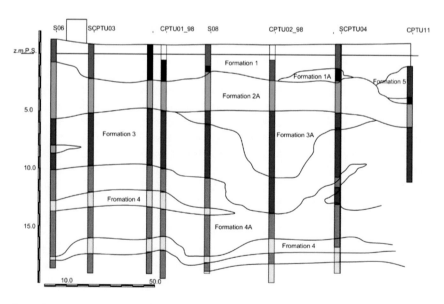

Figure 9.6. Typical stratigraphic section of San Marco's Island (along the section AA' indicated in Figure 9.5).

Table 9.2 Description of soil layers in San Marco's area

Formation	Description	Hydraulic conductivity (m/s)
1	Anthropic fill layer. Heterogeneous material. Prevalently sandy material (SP or SP-ML), with inclusions of sandy silt, graveled sand, bricks, shell remains, and wood remains	2.4×10^{-4} to 2.2×10^{-6}
1A	Anthropic fill layer. Prevalently silty heterogeneous material (ML), with inclusions of shells, wood, and vegetal fibers; pebbles and bricks	1.0×10^{-6} to 5.3×10^{-8}
2	Silt (ML) and sandy silt (ML-SP) with vegetal fibers, peat traces, and shells	2.4×10^{-7} to 2.0×10^{-8}
2A	Clayey silt (ML-CL) with vegetal fibers, peat traces, and shells	7.8×10^{-8} to 5.3×10^{-9}
3	Fine sand (SP)	3.4×10^{-6} to 2.2×10^{-6}
3A	Silty sand (SP-SM) with clayey silt lenses (ML-CL)	2.9×10^{-6} to 2.0×10^{-7}
4	Silty sand (SM) and sandy silt (ML-SP)	2.6×10^{-6} to 1.3×10^{-6}
4A	Clay (CL) with clayey silt lenses (ML-CL)	2.5×10^{-8} to 3.5×10^{-9}
5	Mud from the bottom of the canal	
6	Heterogeneous formation: sand, silt, and clay	

9.4 TIDAL-INDUCED POREWATER PRESSURES IN SOIL

During the site investigation campaign carried out in 1997–1998, 14 piezometers were installed in standpipes within the shallowest anthropic layer and monitored for a short period of time. This first monitoring campaign showed that the hydraulic response is highly variable from site to site, and pore pressure oscillation is attenuated and delayed with respect to tidal oscillation; this suggested that several mechanisms, e.g., distance and state of conservation of gatoli and quay walls, sea water level, and rainfall concur in the pore pressure response of the soil.

In order to better understand the relationship between the pore pressure and the seawater level, nine Casagrande piezometers were installed in 2018 in a pilot site to the south of San Marco's Basilica. Four piezometers are located in the anthropic layer at a depth between 2.0 m and 2.3 m (group A), four piezometers are located in the less permeable formation 2 at a depth between 3.2 m and 3.8 m (group B), and one piezometer is located in formation 3 at a depth of 7.6 m (piezometer C). The Casagrande cell was instrumented with a pressure transducer collecting readings every six minutes and transferring the data to the server, where they could be easily inspected and downloaded for elaboration.

Figure 9.7a shows the oscillation of the average piezometric level measured in soil formations 1–3 over time. The amplitude of the pore pressure oscillation is reduced compared to the tidal wave in all layers, but it reached its maximum in soil formation 2, which is the least permeable and it reached its minimum in soil formation 1, which is the most permeable. During high tide, higher water levels are observed in formation 1 compared to formation 3, thus the hydraulic gradient is directed downward (Figure 9.7b). This proves that during high tides seepage flow from deeper layers could be excluded. During low tide levels, lower groundwater levels are observed in formation 1 with respect to formation 3, and therefore, the hydraulic gradient is directed upward.

Piezometers in formation 1 show a rapid response to tidal oscillation for seawater levels above 0.30 m s.l.P.S. and a much slower response below this threshold. Therefore, maximum levels are close to sea level, whereas the minimum ones stay significantly higher than sea level. This behavior does not characterize soil formation 2 and 3, and it is due to the interaction with the permeable drainage network.

In order to investigate in more detail the pore pressure oscillations in the subsoil, 25 new piezometers were installed in 2019 on 10 monitoring sites distributed in the whole of Piazza San Marco. According to the local conditions, it was decided to use two or three piezometers for each vertical, measuring the absolute pore water pressure in the different formations (1, 2, or 3). The head of all the piezometers was capped and sealed to prevent any contact with the surrounding environment in

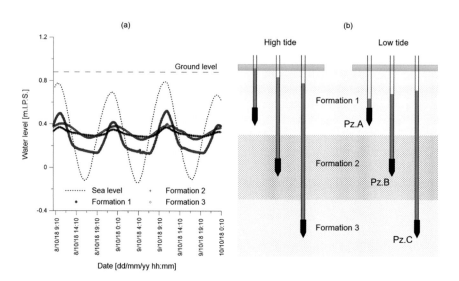

Figure 9.7 (a) Piezometric level and (b) sea water level. Representation of piezometric head in different formations during high and low tides.

order to prevent the submerging tide from affecting the water pressure readings.

Local soil heterogeneities, distance from perimeter quay walls, and the gatoli network, as well as their geometric features and state of preservation, influence the hydraulic response of the entire subsoil system significantly.

A typical pore pressure response in formation 1 as a function of sea level and hourly rainfall is shown in Figure 9.8a for piezometer S08-A. The ground level at this location is +0.86 m, and the area is frequently flooded during high tide.

When the tide rises, water enters the gatoli and can infiltrate the soil through their permeable walls, as shown in phase I of Figure 9.8b. A certain amount of rainfall can also flow through the fissures in the permeable pedestrian paving made of rectangular 0.20 m thick gray stones. When the paving becomes submerged, i.e., in phase II, pressure response is extremely rapid because a larger seepage from the top boundary may occur. This quick response was also observed in salt marshes subjected to submerging tides by Cola et al. (2008), already presented in Chapter 7. The maximum piezometric level may eventually be higher than the maximum sea level, as a consequence of some other significant contributions, such as rainfall or anthropic sources.

When the tidal level decreases, the pressure remains higher than the sea level because of the slow drainage response. Initially, water drains out mainly from the walls of the *gatoli* (phase III), and then, at very low tide level, only from quay walls or very deep ducts, as in phase IV, showing a

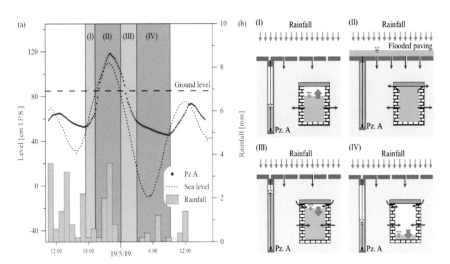

Figure 9.8 (a) Sea level, piezometric level (S08) and hourly precipitation (courtesy of Centro Previsioni e Segnalazioni Maree, Venezia) on 19 May 2019. (b) Schematic representation of water flow in phases I–IV indicated in subfigure (a) (Ceccato, Simonini, and Zarattini 2021).

slower pressure dissipation rate. In phase IV, the soil may reach partially saturated conditions, thus reducing its permeability.

In other words, the local vertical component of the hydraulic gradient between the two layers is directed downward for higher tides and upward for lower ones. Thus, infiltration from the deeper layer is impossible during high tide peaks. In the shallower layer, maximum water levels are close to sea level, whereas the minimum ones are higher; moreover, pore pressure may stay higher than sea level during tide decrease. This may be a key phenomenon that should be considered when carrying out uplift analyses of architectural elements, as well as when determining the long-term stability of paving, especially in the lowest area of San Marco's square.

REFERENCES

Bettiol, G., Ceccato, F., Pigouni, A. E., Modena, C. and Simonini, P. 2016. Effect on the structure in elevation of wood deterioration on small-pile foundation: Numerical analyses. International Journal of Architectural Heritage 10(1), 44–54. https://doi.org/10.1080/15583058.2014.951794.

Bortoletto, M. 2019. Relazione archeologica – Interventi di salvaguardia dell'Insula di Piazza San Marco a Venezia (in Italian).

Ceccato, F., Simonini, P., Köppl, C., Schweiger, H. F. and Tschuchnigg, F. 2014. FE analysis of degradation effect on the wooden foundations in Venice. Rivista Italiana di Geotecnica 2(2), 27–37.

Ceccato, F., Simonini, P. and Zarattini, F. 2021. Monitoring and modeling tidally induced pore-pressure oscillations in the soil of San Marco's Square in Venice, Italy. Journal of Geotechnical and Geoenvironmental Engineering 147(5), 1–14. https://doi.org/10.1061/(ASCE)GT.1943-5606.0002474.

Cola, S., Sanavia, L., Simonini, P., and Schrefler, B. 2008. Coupled thermohydromechanical analysis of Venice lagoonsalt marshes. Water Resour. Res., 44, W00C05. doi:10.1029/2007WR006570

Fletcher, C. A. and Spencer, T. (Eds.). 2005. Flooding and Environmental Challenges for Venice and its Lagoon: State of Knowledge. Cambridge: Cambridge University Press.

Ruol, P., Favaretto, C., Volpato, M. and Martinelli, L. 2020. Flooding of Piazza San Marco(Venice): Physical model tests to evaluate the overtopping discharge. Water 12(2), 427. https://doi.org/10.3390/w12020427.

Salandin, P. 2020. Evaluation of flooding of the San Marco's Island. In Analysis of Possible Interventions to Safeguard the San Marco's Island from High Tides Part 3. Padova: University of Padua.

Volpato, M. 2019. Relazione idrologica e idraulica—Interventi di salvaguardia dell'Insula di Piazza San Marco a Venezia. Venice, Italy: Consorzio Venezia Nuova. (in Italian).

Chapter 10

Concluding remarks

The contents of this book constitute a summary of multiple studies that I, together with colleagues and experts from various scientific fields, have carried out over the last 20 years; this book therefore represents the results of my personal experiences and points of view.

From many chapters of the book, it clearly emerges that both the city of Venice and the lagoon's ecosystem suffer from environmental degradation due to the effects of subsidence and eustasy combined with anthropic pressure, all of which increased significantly over the second part of the last century.

More particularly, the eustatic sea level rise is the direct consequence of global climate change, which, in Northern Italy, is also characterized by an increasing number of storms and heavy, concentrated rainfall, coupled with more frequent high-tide surges in the lagoon and strong winds. The evidence of climate change can be clearly and easily confirmed by counting the number of exceptional tides, which has rapidly increased over the last few decades.

Therefore, if a friend asks me whether Venice is sinking due to soft soil conditions, or what should be done in order to save the city, I typically reply that from a geotechnical point of view, the conditions of the subsoil are not terrible and subsidence is not on the brink of worsening, since the impact of natural subsidence is rather limited; otherwise, Venice would have already sunk.

The question that arises is related to the long-term effectiveness of the defense systems against flooding, which, as already pointed out, is mostly due to eustatic sea level rise.

Throughout the centuries, the Venetians have always protected their territory, for strategic reasons and for the survival of the population.

After the terrible flood of 1966, several colossal interventions were planned and completed: the MoSE mobile barrier system, elevation of the islands' pavements, reinforcement and extension of breakwaters, coastal nourishment, and erosion mitigation works. All these interventions have been designed using a multidisciplinary approach, including both engineering

DOI: 10.1201/9781003195313-10

and life sciences, considering possible future scenarios, especially related to the estimated increase in sea level.

If the climate change trend cannot be modified through a worldwide agreement among countries concerning the reduction of the anthropic pressure on the Earth, the interventions set up in recent decades may not be enough to preserve the delicate ecosystem that forms the landscape of the city and of the surrounding lagoon over the long term.

Nevertheless, all these interventions have required and continue to require accurate knowledge and modeling of the geotechnical behavior of the Venetian soils, mostly composed of non-plastic silt, combined with clay and sand in an erratic and heterogeneous manner.

The Venetian heterogeneous silt, showing a common mineralogical origin through the degradation of the original sands, is a soil whose particles essentially interact with each other through mechanical rather than electrochemical forces. Except for some organic layers, Venetian clays are low-activity materials, prevalently mixed with silt and sand, and only slightly impact the overall subsoil behavior. This quality of the soil has made it possible to describe its mechanical behavior by considering a unified approach to estimate relevant properties, on which constitutive models are based.

The experimental test sites at Malamocco and Treporti provided a unique opportunity to give deep insight into the behavior of these poorly structured soils, extremely sensitive to stress relief due to sampling. The evaluation of stress history as well as the current stiffness, fundamental key parameters to estimate settlements over highly heterogeneous soil profiles, has been properly measured only through the large load test at Treporti, without which any accurate mechanical description of the Venetian soil would have been impossible. The effectiveness of the scientific findings certainly helped to design the foundations of the massive MoSE tilting gate system, which offers a costly maintenance but efficient solution to protect the city against exceptional tides. The MoSE system has recently been tested and has proven to be very effective at defending the city and lagoon from exceptional tides.

In addition, the large load test results have been properly used to calibrate site-specific geotechnical tests, such as the piezocone and the dilatometer. These devices have been demonstrated to represent applicable tools, estimate relevant soil properties, and calculate long-term displacement of different types of structures in the entire lagoon.

Geotechnical engineering plays a relevant role also in the preservation of the magnificent urban landscape as well as the unique lagoon ecosystem.

It is particularly remarkable that the long-term stability of the numerous historical buildings, bell towers, and quay walls entirely rely on short and closely spaced wooden piles, which have been proven to be particularly effective as soil-improving elements. What the Venetian citizens of the past have been able to design and build is incredible, on islands characterized by extremely difficult ground conditions and daily tide oscillations, by using any type of innovative geotechnical techniques, such as soil dredging, excavation protection, and draining, pile driving, and so on, many centuries before the principles of soil mechanics had been established.

An example is given by the 18th-century drainage network installed in Piazza San Marco, which currently needs renovation. After comparing different solutions, the one selected by contemporary designers for the preservation of one of the most famous city squares in the world was merely an adjustment and local improvement of the already existing buried network of drainage channels built centuries ago. This choice derived from the geotechnical analysis of subsoil stress equilibrium and strain compatibility under oscillating tides, carried out by applying concepts learned a few years before, through observation and modeling of the coupled mechanical response of salt marshes in unsaturated conditions, to Piazza San Marco. I am extremely proud of this contribution, which has allowed the selection of a suitable and relatively cheap solution for the preservation of Piazza San Marco.

Finally, from my personal experience, I would like to conclude by stating that the historic city of Venice and its lagoon can be perceived as an enormous open-air laboratory in which geotechnics, in a broader context of interdisciplinary exchange, can learn lessons from the past and play a fundamental role, today and in the future, in studying and modeling different natural and anthropic processes in order to preserve such a magnificent natural and cultural heritage.

Index